Green Wisdom

GREEN
WISDOM

ARTHUR W. GALSTON

Illustrated by Lauren Brown

cop b

Basic Books, Inc., Publishers   *New York*

Most of the text appeared originally in *Natural History*

Library of Congress Cataloging in Publication Data

Galston, Arthur William, 1920-
  Green wisdom.

  Bibliography
  Includes index.
    1. Botany—Addresses, essays, lectures.
  2. Botany, Economic—Addresses, essays, lectures.
  I. Title.
QK81.G18   580   80-68183
ISBN: 0-465-02712-1

To two great teachers

Loren Petry

who struck the first botanical spark within me and

Harry Fuller

who helped keep the spark alive

CONTENTS

LIST OF ILLUSTRATIONS   ix

PREFACE   xi

Introduction
The World of Green and the World of Man   3

Making a Living Can Be Tough

1 New Ways to Increase Man's Food   13
2 The Water Fern–Rice Connection   18
3 The Prodigal Leaf   24
4 The Membrane Barrier   31
5 The Blind Staggers   39

The Plant Coordinates Itself

6   Botanist Charles Darwin   49
7   Which End Is Up?   59
8   Rotten Apples and Ripe Bananas   67
9   Turning Plants Off and On   74
10 Sex and the Soybean   80
11 Plants Have a Few Tricks, Too   89

Contents

The Plant Moves About

12 The Language of the Leaves   97
13 A Basic Unity of Life   104
14 The Limits of Plant Power   109

Life, Death, Immortality, and Other Problems

15 The Immortal Carrot   117
16 The Naked Cell   126
17 Molding New Plants   134
18 Here Come the Clones   140
19 Plant Cancer   147
20 In Search of the Antiaging Cocktail   155

Offbeat Plants

21 A Living Fossil   165
22 Guayule Bounces Back   171

Plants and the Environment

23 How Safe Should Safe Be?   181
24 The Organic Gardener and Anti-intellectualism   188
25 Coda   194

BIBLIOGRAPHIY   199

INDEX   203

LIST OF ILLUSTRATIONS

The green plant as a solar machine 4

The solar thermonuclear furnace 5

The rice plant and *Azolla*, the water fern 20

Runty rice plants and sturdy ones 21

Leaf stomata 27

Ferocactus, the spiny water saver 30

Two plants that cause the "blind staggers" 40

The wild cucumber, studied by Asa Gray 54

Seedlings turn toward the light 56

How Swedish ivy responds to light 57

Roots turn down and stems turn up 60

How a plant orients itself 62

Reaction wood in a pine tree 65

A rotten apple spoils the barrel 68

The winter growth of trees 76

A mobile sex hormone in plants 84–85

The hawthorn and the desert *Acacia* 91

What makes plants open and close 100

Albizzia leaves, open and closed 102

The immortal carrot in the laboratory 121

The natural life cycle of the carrot 122–23

The "pomato" 132

(ix)

List of Illustrations

A witches–broom on hackberry   150

A stem of sunflower gets a tumor   152–53

Kakabekia, highly enlarged   166

A plant of guayule   172

PREFACE

THERE IS a functional wisdom built into all living things; without it they could not survive in a world of constant challenge and danger. This kind of wisdom is easy to recognize in animals, to whom we frequently attribute our own patterns of thought and rational decision-making. The camouflage of a butterfly, the burrowing of a mole, the evasive hopping of a rabbit, the southward migration of a bird as winter approaches—all these make sense to us as logical tactics adopted by successful creatures seeking to escape danger. But in truth, none of these behavioral patterns results from the organism's considered, rational approach to a problem. Rather, each follows automatically from interaction of an external stimulus with built-in mechanisms of chemistry, structure and physiology. The apparently wise southward migration of birds as a northern-hemisphere winter approaches is an automatic result of the creature's perception of the shortening of the day, which is in turn the automatic result of the earth's movement around the sun. The bird can be fooled into migrating southward in midsummer by confinement in a chamber with artificially shortened daylengths, or prevented from migrating in the fall by artificially lengthened days. In the latter instance, the bird would probably freeze or starve to death over the winter. Thus, the real basis for the bird's "wise" behavior is the evolutionary development of response patterns that favor its survival when challenged by potentially adverse environmental forces.

Preface

Green plants have this kind of wisdom, too. Indeed, their continued survival on earth several billions of years after the appearance of the first green cell is all the evidence one needs to believe that fact. But many people, even those impressed by the elegance and subtlety of survival mechanisms built into all creatures, are unaware of the remarkable behavioral strategies built into plants. I hope these brief essays, most of which appeared as monthly columns in *Natural History*, will increase the reader's appreciation of what I like to think of as Green Wisdom.

ARTHUR W. GALSTON

New Haven, Connecticut
September 1980

Introduction

The World of Green
and the World of Man

IT IS A PARADOX that the word "plant" should have come to mean both a green organism growing quietly in the sun and a noisy factory consuming fuel and discharging smoke as it turns out its product. These two kinds of plants, so different in form and function, are linked by more than their common name. The leaves of green plants, products of several billion years of evolution and still imperfectly understood by man, are unique in the biological world for their ability to capture and store light energy in the process of photosynthesis. Using solar energy, they combine carbon dioxide and water into organic molecules like sugars, in which the physical energy of the sun's radiations has been transformed into the energy of chemical bonds. Man-made plants, composed of brick and metal, can function only by reversing that process, liberating through oxidation the energy trapped in organic molecules. Historically, the energy used by manufacturing plants has come from green plant products: wood, coal, and oil. Thus, factory has

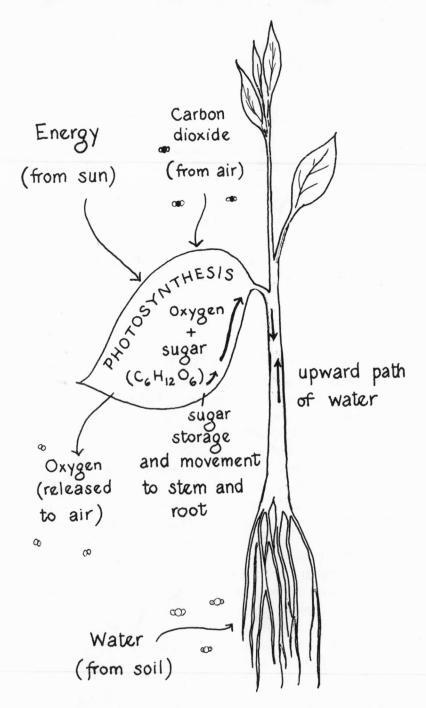

Energy
(from sun)

Carbon
dioxide
(from air)

PHOTOSYNTHESIS

Oxygen
+
sugar
$(C_6H_{12}O_6)$

Oxygen
(released
to air)

sugar
storage
and movement
to stem and
root

upward path
of water

Water
(from soil)

The green plant harvests the sun's energy, storing it in the form of the
chemical bonds of sugars and other compounds.

4 hydrogen atoms,
each with an atomic
number of 1 and
a mass of 1.008
(total mass = 4 × 1.008 = 4.032)

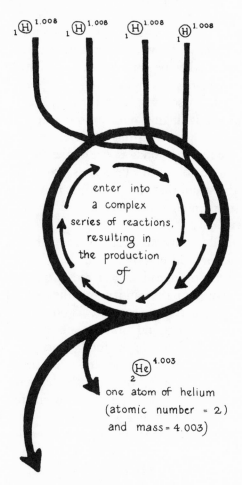

$_1$(H)$^{1.008}$ $_1$(H)$^{1.008}$ $_1$(H)$^{1.008}$ $_1$(H)$^{1.008}$

enter into
a complex
series of reactions,
resulting in
the production
of

$_2$(He)$^{4.003}$

one atom of helium
(atomic number = 2)
and mass = 4.003)

and the release of
large amounts of
energy $(E = mc^2)$
equivalent to the
difference in mass
(4.032 - 4.003 = .029)

The sun is a giant thermonuclear reactor
placed safely 93,000,000 miles away from
the earth. It works by fusing hydrogen nuclei into helium. This is the same reaction
physicists hope to be able to control on
earth.

fed on field, and plant has fed on plant. Only recently, in the atomic age, have alternative energy sources appeared to break this otherwise compulsory energy coupling.

To the physicist, energy is simply the capacity to do work. Easily utilizable energy is concentrated in molecules called fuels; treated in appropriate ways, fuels release their contained energy, which can then be put to work. For example, when heated to their kindling temperature in the presence of oxygen, fuels enter a self-sustained burning reaction. As the electronic bonds linking one carbon atom to another are broken, and their electrons transferred to oxygen, the energy stored in the bonds is released as heat. This heat can then be used to warm our homes, to cook our food, to produce the steam that turns turbines of electrical generators, to run engines or vessels, and to do thousands of other useful jobs.

While wood, coal, and oil are all derived from green plants, they are valued very differently by modern man. Coal and oil, having been produced over large spans of geological time and present in finite quantity over the earth, are essentially nonrenewable energy sources. Wood, being constantly produced as trees and other plants grow, can be renewed and replenished; so can the food plants that we eat. Since edible meat and fish are produced when animals consume vegetation or other smaller animals that eat vegetation, the solar-powered green plant is the ultimate source for virtually every calorie of energy we use.

The sun is a thermonuclear device in which hydrogen atoms are fused, in a complex cycle of reactions carried out under very high temperature and pressure, to produce helium. Why should this thermonuclear fusion of hydrogen to helium in the sun liberate so much energy? The answer comes from the interconvertibility of matter and energy, as expressed by the now familiar Einstein equation, $E = mc^2$. This tells us that the energy content (E) of a bit of matter is equal to its mass (m) multiplied by the square of the velocity of light (c^2). Each of the four hydrogen nuclei entering the solar furnace has 1.008 units of mass, while the helium atom resulting from their fusion has 4.023 units of mass.* Thus, the net effect of the reaction $4H \rightarrow He$ is to destroy 0.009 units of mass [$4 \times 1.008 = 4.032$; $4.032 - 4.023 = 0.009$], or about 0.22 percent of all the mass entering the reaction. This "destroyed" mass appears as energy: thus, 0.009 grams of mass would become equal to $(0.009) \times (3 \times 10^{10}$ centimeters per second$)^2$ ergs. This is equal to 0.081×10^{20} ergs or roughly 2×10^{11} calories. Put in more conventional terms for nonmetrical Americans, about 0.0003 ounces of matter would yield, in the nuclear furnace, some two hundred billion calories of energy. No wonder the proponents of nuclear energy are so enthusiastic about its possibilities.

Ever since the accident at the Three Mile Island nuclear power plant, however, the debate about nuclear versus conventional energy has become more strident, and nuclear energy supporters have become a bit more conservative. Since all mechanical devices are to a certain extent imperfect, even the slight possibility of a nuclear accident near a city becomes to many an intolerable risk. And as nuclear plants proliferate, even the low probability of an accident of some sort becomes a virtual certainty, given enough time.

Those who call for maximum safety in energy devices frequently point to solar energy as the way to go. Not

---

*Another way to look at this is to consider Avogadro's number of atoms (about 6×10^{23}). This is the number of atoms of a particular type it takes to weigh one atomic weight of that element in grams. Thus, Avogadro's number of hydrogen atoms weighs 1.008 grams; the same number of helium atoms weighs 4.023 grams.

only do solar energy gatherers and transducers work without obvious risk, the solar energy on which they draw is itself virtually inexhaustible. But as we have seen, solar energy is itself the result of a thermonuclear reaction. What makes the sun safe for us on earth is that its nuclear reactions occur 93 million miles away, out in space. This distance insures that most of the potentially harmful radiations resulting from the reactions will never reach us, either because they are absorbed by matter in space between the sun and us, or because even the 500 seconds it takes electromagnetic radiations to reach us from the sun is longer than the lifetime of unstable particles whose decay gives rise to radiations.

The green plant is thus a self-sustaining machine that converts safe solar energy into a multiplicity of forms usable by man. The key to this process in higher plants is the *chloroplast*, a disc-shaped body about the size of one of our red blood cells. Each chloroplast contains an assembly of pigments, including green chlorophyll, arranged in stacks of membranes that are piled up like poker chips. These piles are then attached to other long, flat membranes. In this highly structured array, which is an elegantly microminiaturized version of the cruder solar-energy capture devices designed by physicists, light energy is converted into chemical energy.

Although this process is relatively inefficient (about 2 percent of the energy falling on plants is conserved, according to a recent estimate), it still results in a massive fixation of carbon over the surface of the earth. A generally accepted estimate of the annual total photosynthetic carbon fixation is 200 billion tons per year, a figure that dwarfs the total of man's activities. About half of the total carbon is fixed on land, half in the waters of the earth.

The World of Green and the World of Man

While intensive agriculture tries to maximize photosynthetic productivity over a small fraction of the earth's land surface, much remains to be done. Since the world's population of more than 4 billion people continues to increase at a rate of about 1.8 percent per year (approximately 72 million new mouths to feed in 1980), there can be no more urgent priority on mankind's agenda. Our food resources are already strained, and just as a shortage of energy for industrial purposes is creating much international tension, so a continued imbalance between food production and population growth could result in catastrophe. The green plant can help stave off both these crises, but it cannot be forced to do so. It must be studied, cajoled, even seduced into revealing its secrets. The practical benefits of this exercise for mankind may be great . . . or zero. The wonder is that the random, unplanned inquiries of so many investigators have yielded so much, and that the support of high quality basic research has proved to be the best way to accomplish so many practical objectives.

Making a Living

Can Be Tough

1

New Ways to Increase
Man's Food

THE FARM CROPS of the world differ tremendously in their rates of growth and total yield per acre. Grasses of tropical origin, like sugarcane, sorghum, and maize, have the highest recorded yields, frequently more than twice that of such other important crop plants as soybeans. Why the difference? In a world in which food production has difficulty keeping pace with the needs of increasing numbers of hungry people, the question is of more than theoretical interest. Some exciting clues have been uncovered in the laboratory in the last few years that may ultimately lead to improved plant productivity in the field.

The two basic processes to consider are photosynthesis and respiration. In photosynthesis, light energy captured by chlorophyll-containing bodies in the cells of green plants is used to form sugar and oxygen out of carbon dioxide from the air and water from the soil. The immediate photosynthetic product, sugar, contains the absorbed energy in a stored chemical form. Some of the sugar is subsequently transformed into the other organic materials in the plant, such as starch, cellulose, fat, protein, and vitamins. These various materials, when consumed by animals, become both the energy source and the basic building blocks for all higher creatures, including man.

In respiration, the energy bound up in the sugar molecule is released. This process, common to all cells—plant and animal alike—involves the oxidation of sugar to yield carbon dioxide, water, and energy. Respiration is essentially the reverse of photosynthesis, although it takes place in all cells continuously, day and night, whereas photosynthesis proceeds only in green cells and only in the light.

The productivity of a plant can be measured by comparing the amount of photosynthesis, by which it creates organic matter, and its total amount of respiration. For a plant to grow rapidly, it must capture much more energy during periods of light than it uses for respiration in light and darkness.

In the presence of light, plants release much less carbon dioxide through respiration than they absorb by photosynthesis. Under optimal conditions, the rate of photosynthesis can be up to twenty times as rapid as the rate of respiration in a green cell. The standard laboratory technique for estimating the true rate of photosynthesis in a green cell has been to measure the rate of carbon dioxide fixation in the light and then to correct that figure for carbon dioxide lost by respiration in the dark.

That simple correction, although valid logically, turns out to be biologically misleading. It assumes that the respiratory rate as measured in darkness is the same as the respiratory rate in the light. About fifteen years ago, however, it was noticed that when laboratory plants in the dark are suddenly exposed to light, they release a gush of carbon dioxide, obviously from some chemically bound form of the gas in the leaf. This process involving the extra yield of

carbon dioxide in the light was named *photorespiration.*
Agricultural researchers became interested in the process
when they discovered that photorespiration can release as
much as half the carbon dioxide fixed by a plant. Such
photorespiration was undoing the beneficial effects of
photosynthesis. Furthermore, it was learned that high-
yielding plants tend to have low rates of photorespiration,
and vice versa. This suggested that photorespiration wastes
a plant's energy and that if the process could be modified
in an inefficient crop like soybeans the rate of carbon dioxide fixation, and therefore ultimately the rate of crop
growth, could be increased. This has turned out to be more
difficult to accomplish than to theorize about, but significant progress has been made.

The first important question, "What is the source of the
carbon dioxide involved in the photorespiratory gush?" was
answered relatively quickly. Conventional tracer experiments with radioactively labeled carbon showed that
much, if not all, of the released carbon dioxide came from
the oxidation of a simple substance called *glycolic acid.*
Glycolic acid oxidation probably does not occur in significant amounts during normal respiration in darkness. When
a light is turned on, however, the process begins, and its
release of carbon dioxide partly negates the photosynthetic
fixation of the gas, which also starts when the light is
turned on. Must these two light processes—one favorable
to plant growth and the other apparently detrimental—
necessarily go on together? Is there some possibility that
photorespiration might be slowed down or eliminated without harming the plant?

One approach to deleting or diminishing the reaction
would be to use genetic techniques to select and breed
strains of plants that have low photorespiration rates; another would be to find a chemical substance that would inhibit glycolic acid oxidation without affecting the rest of

the plant's respiration. (Respiration is, after all, required for continued energy release; it would be counterproductive to stop it entirely.) Both approaches have been tried with partial success. In particular, an inhibitor called alpha hydroxy pyridine methanesulfonic acid (HPMS for short) substantially decreases glycolic acid oxidation when sprayed on plants. HPMS also causes glycolic acid to accumulate in the plant. This glycolic acid buildup is ultimately converted to stored plant materials with the overall effect that carbon dioxide fixation in the light is increased two- to fourfold. These results have so far been accomplished only in the laboratory, but it is believed that they can be matched in the field to increase plant yield.

The environment can also be manipulated to lower the rate of photorespiration. At the moment, such methods are laboratory tricks, not applicable to field agriculture, but they might be effective with crops currently grown in greenhouses and, in the future, used for crops raised under plastic bubbles. One laboratory technique employs temperature moderation. Photosynthetic carbon dioxide fixation in a plant like corn, which has a very low rate of photorespiration, increases progressively when the temperature is raised to about 97° F. On the other hand, carbon dioxide fixation in wheat, which has an active photorespiration rate, may actually decrease when the temperature is raised. These opposite effects occur because the rate of photorespiration accelerates more with a rise in temperature than does the rate of photosynthesis. Thus, raising the temperature of some plants can induce higher yields, whereas lowering the temperature of others induces a parallel result. This fact may account to some extent for variations in the efficiency of different plants at different latitudes and could possibly be used for crop control in the future.

Another approach stems from the observation that the photorespiration rate mounts progressively as the amount of available oxygen increases all the way up to an oxygen concentration of almost 100 percent in the air surrounding a plant. Since photosynthesis is not similarly affected, net carbon dioxide fixation and crop yield decline as oxygen concentration rises. This suggests that decreasing the oxygen content below the 20 percent level found in ordinary air might decrease photorespiration and improve the growth of some plants. This actually works out in practice. Soybean plants, for example, grow much more lushly in an atmosphere of 5 percent oxygen than they do at 20 percent, and they produce the same number of flowers and pods. But so far an unexpected result has precluded a better harvest. When the pods are opened up, the seeds—the most valuable part of the harvest—are found to have aborted or failed to fill out. It appears that some system important to seed development requires the higher oxygen concentrations. To realize the benefits of this experiment in plant growth, the oxygen level should therefore be lowered only during the vegetative or preflowering stage. Once flowering has occurred, oxygen concentration may have to be raised to atmospheric levels for normal seed development.

These and other experiments with photorespiration have introduced us to some of the complexities of plant energetics and growth. Future efforts will lead us to more theoretical knowledge about these processes and may point to methods for increasing food for mankind. Undoubtedly such frustrations as the stunted soybean seeds will recur, but some of the many experiments under way may show better results. At this point, prospects are hopeful that genetics may come to the rescue, through the selection and breeding of highly productive varieties with low photorespiration rates.

2

The Water Fern–Rice Connection

THE WATER FERN *Azolla* is usually studied by first-year botany students and professional botanists, but it is practically unknown to almost everyone else in the Western world. Together with two related genera, it belongs to the small group of ferns adapted to life in an aqueous environment. This makes *Azolla* interesting from an ecological and evolutionary point of view, but its small size, limited distribution, and apparent lack of economic importance have contrived to keep it obscure. For a surprising reason learned during a recent trip to the Far East, I believe this plant, commonly called the mosquito fern, is soon likely to become much better known and intensively studied outside the Orient.

Colonies of *Azolla* plants, each perhaps a centimeter in diameter, float on bodies of fresh water, their flat leaves arranged in two alternate rows for buoyancy. Each leaf has two parts: an upper green lobe—with stomata, or orifices—which is active in photosynthetic production of food, and a lower colorless lobe, which is probably useful in the absorption of water and minerals. Short roots, which extend from the junctions of the branches a few millimeters below the floating frond, also serve as organs of absorption, although their orientation may make them additionally useful as a keel for stabilizing the floating plant body. Reproduction is mainly vegetative—new fronds develop from buds that separate from the mother frond as the connecting stems break up.

For as long as men can remember, the peasants in several villages of Thai Binh province, the most intensively cultivated area in North Vietnam, have produced extraordinarily high yields of rice. The farmers attribute this bounty to the joint culture of rice and the *Azolla pinnata* fern in their paddies. In the winter season, when rice seedlings are transplanted, small starter colonies of the fern are placed in the paddies. They multiply rapidly by vegetative propagation and within one or two months completely cover the surface of the water. Growing together, the fern and the rice make the green paddies efficient absorbers of solar radiation.

An abundant growth of the fern is invariably accompanied by rice yields 50 to 100 percent greater than that achieved in adjoining paddies that are not "seeded" with *Azolla*. The relationship between the fern and the rice seems hard to understand until one realizes that the leaves of the water fern contain little pockets in which colonies of a particular strain of the blue-green alga called *Anabaena* grow. This alga not only fixes carbon dioxide into sugars by photosynthesis, it also fixes nitrogen from the air into forms that can be utilized by rice and are therefore of great value to the growth of the cereal.

It has long been known that *Anabaena* and two other blue-green algae found living free in the waters of rice pad-

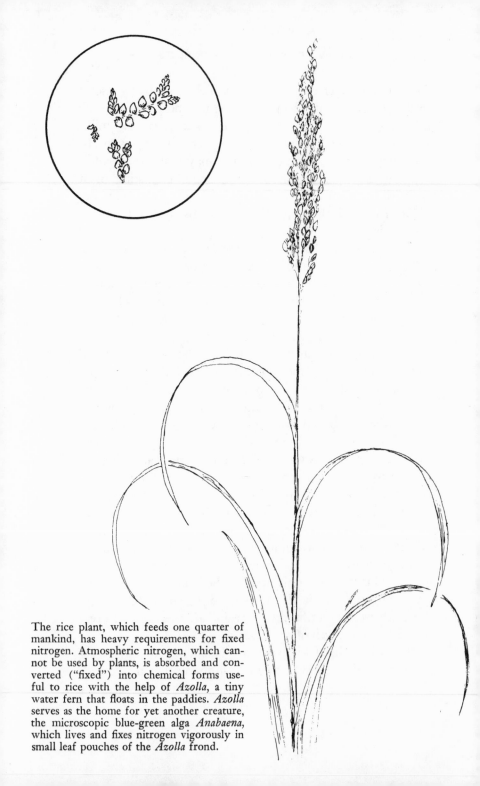

The rice plant, which feeds one quarter of mankind, has heavy requirements for fixed nitrogen. Atmospheric nitrogen, which cannot be used by plants, is absorbed and converted ("fixed") into chemical forms useful to rice with the help of *Azolla*, a tiny water fern that floats in the paddies. *Azolla* serves as the home for yet another creature, the microscopic blue-green alga *Anabaena*, which lives and fixes nitrogen vigorously in small leaf pouches of the *Azolla* frond.

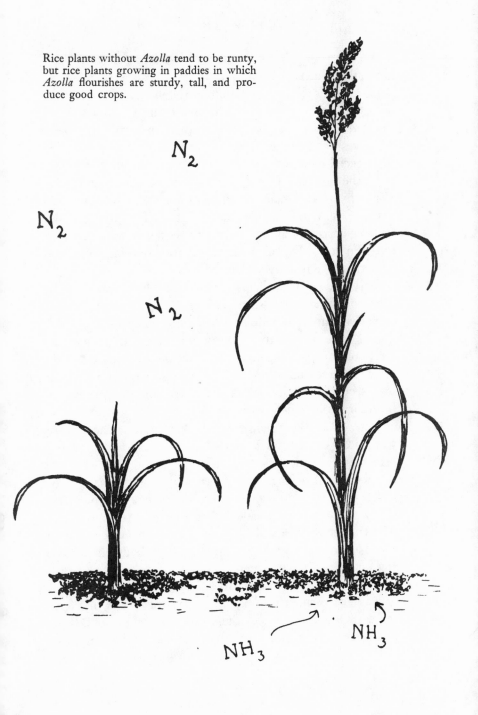

Rice plants without *Azolla* tend to be runty, but rice plants growing in paddies in which *Azolla* flourishes are sturdy, tall, and produce good crops.

dies fix nitrogen, but the *Anabaena-Azolla*-rice mutualism has only recently been uncovered. It appears that the *Anabaena–Azolla* combination is much more efficient in fixing nitrogen than any of the free-living forms alone, although no one knows why; furthermore, the *Anabaena* that lives in the water fern's pocket is uniquely adapted to that special environment and will not grow well if removed from the pocket and cultured in the water of the paddy. Why this is so remains a mystery, but it can be presumed that the symbiotic alga strain has lost its ability to make some essential compound that is furnished in the pocket of the fern leaf. The alga, in turn, probably improves the growth of the fern by virtue of the nitrogen it fixes.

Only one or two villages in Thai Binh have traditionally been able to furnish "seed" colonies of water fern for inoculation into rice paddies during the winter transplantation. The fern tends to die away during the late spring and early summer; at that time, the plant goes through a cycle of senescence and yellowing and is also subject to attack by insects that eat its fronds and roots. Somehow, the Thai Binh farmers learned over the years how to protect *Azolla* colonies from these dangers and when winter came, they were able to sell small starter colonies at very inflated prices. So valuable was the water fern considered to be that the secret of protecting it, guarded by a formal set of taboos and restrictions, was passed on in a solemn ceremony only to young males when they entered into independent family life and farming. It was never revealed to the women because they might marry outside the village and take it away with them.

As a result of food shortages during the thirty years of war in Vietnam, the Thai Binh peasants were urged to share their secrets with others. The technique of *Azolla* preservation, it appears, involves regulating the acidity of the seedstock-producing paddy. Under normal conditions, a paddy becomes progressively more alkaline during the rice-growing season, and this favors the senescence and death of the water fern culture. If acidification to carefully prescribed formulas is carried out under conditions of high light intensity at the proper time of year, the cultures tend to survive. But even the surviving fern cultures are vulnerable to insect attack. These are now successfully warded off by an organophosphorus insecticide called Wofatox, imported from East Germany.

Azolla culture is no longer a secret; consequently, use of the fern is spreading rapidly in the rice-growing regions of North Vietnam and in southern China. In the absence of substantial fertilizer production and of the foreign exchange necessary for purchase of such materials abroad, the water fern may prove to be of critical importance in maintaining food production for an expanding Vietnamese population, which now exceeds fifty million in North and South Vietnam combined. But since ferns produce no seeds, *Azolla* culture will probably continue to depend on the careful hand manipulation of fresh starter colonies for some time to come.

3

The Prodigal Leaf

ON THE FACE OF IT, plants are water wastrels. A single corn plant contains about two liters, or slightly more than two quarts, of water, but during the course of its growth, it probably removes about one hundred times that amount from the soil. Most of the water is simply evaporated into the air. An acre of corn would thus use about 1,200 metric tons of water during a growing season, the equivalent of an eleven-inch rainfall. Such a prodigal use of water can be dangerous, even lethal, to plants growing in areas of limited moisture.

Why such low efficiency in the use of water? It has to do with the architecture and function of the .leaf. The main job of the leaf is nutritional, for it is in the leaf that carbon dioxide and water, catalyzed by light energy, interact to form oxygen and various organic compounds in the process of photosynthesis. In this reaction, the energy of the light absorbed by chlorophyll is stored for future use in the form of sugar. The sugar is later oxidized in the process of respiration, which liberates the energy originally stored in the sugar molecules.

In order to perform its job, the leaf has a large flat area

The Prodigal Leaf

(for light absorption), is highly hydrated (leaf cells are about 90 percent water), and is perforated to permit the passage of gases (carbon dioxide entry and oxygen loss). The leaf is thus something like a wet towel hanging on a clothesline: warmed by the sun and fanned by the wind, it necessarily evaporates large quantities of water. The process of water vapor loss from the aerial parts of plants, called transpiration, is one that most plant physiologists consider harmful or, at best, an inevitable consequence of the photosynthetic mechanism. In effect, to get the benefits of photosynthesis, you have to accept the evils of transpiration.

Not all scientists are happy with this concept. Full of admiration for the efficient design of natural systems, they look for unappreciated benefits brought about by transpiration. One obvious possibility is evaporative cooling. A corn plant in the noonday Iowa sun in midsummer receives about 1.5 calories of radiant energy per square centimeter per minute, of which about two-thirds, or one calorie per square centimeter per minute, is absorbed. During a growing season, this amount of energy would be roughly equivalent to that released by burning 200 tons of coal for each acre of corn. Such a prodigious amount of absorbed energy would completely desiccate the plant, were it not that each gram of water evaporated absorbs more than 500 calories. Such a heat loss would occur, on the average, once per hour per 100 square centimeters of leaf surface, and would represent almost 10 percent of all the energy absorbed. When you add to that the energy loss to leaves by reflection, reradiation, and the movement of air, you approach a tolerable heat buildup. Thus, as long as plenty of water is delivered to the leaf, the evaporative cooling produced by transpiration helps keep it from being cooked to death.

Another possible useful function attributed to transpiration is its facilitation of the upward movement of sap in the wood of the stem. For as the leaf blade dries out by transpiration, it pulls water from nearby veins. The many leaves on a plant contrive in this manner to exert a large total "pull" on the continuous water columns in the microscopic tracheids, or tubular cells, and vessels of the woody tissues of the stem. This combined pull is enough to cause sap to rise to the tops of the tallest trees. Since sap contains not only water but also minerals absorbed from the soil and some organic materials produced in the plant, the net effect of transpiration is to speed the upward movement of these dissolved materials from the root to the leaf and ultimately to the tops of trees. In many forests, much of the total amount of available mineral nutrient is found in the trunks of trees. This fact is responsible for the slash-and-burn agriculture practiced by some nomadic peoples as a means of restoring minerals to the soil for crop plants.

The plant is not totally helpless in resisting water loss. For one thing, almost all leaves are covered by a waxy, water-impermeable layer called the cuticle, which overlays the epidermis. Water loss is minimal through most cuticles. The bulk of transpiration occurs through stomata, minute openings on the upper and lower epidermis of leaves and some stems. These intercellular pores are bounded by specialized, sausage-shaped epidermal cells, called guard cells, whose changing form determines the degree to which the pore is open. When the guard cells are turgid, or swollen with water, they become crescent shaped; they then touch only at their ends, and the pore between them is fully open. When the guard cells lose water and turn flaccid, they become linear, and the pore is partly or completely closed. These changes in shape are due to the differential thickness of the guard cell walls. Their outer surfaces, away from the pore, are thin. It therefore stands to reason that as the pressure of fluid, or turgor, within the guard cells increases, the thinner outer walls will balloon out, carrying along the inner walls and opening the pore. When these guard cells lose their turgidity, the elastic strength of the thicker inner walls pulls the guard cells back into a linear orientation, and the pore closes again.

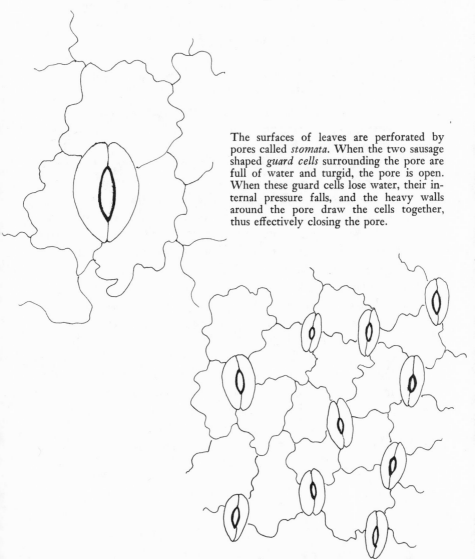

The surfaces of leaves are perforated by pores called *stomata*. When the two sausage shaped *guard cells* surrounding the pore are full of water and turgid, the pore is open. When these guard cells lose water, their internal pressure falls, and the heavy walls around the pore draw the cells together, thus effectively closing the pore.

A corn leaf has about 100,000 stomata per square inch of leaf surface, or about 15,500 per square centimeter. The upper surface of a typical leaf has about 20 percent more stomata than the lower surface per unit area, but both the number and ratio vary with the plant and with the conditions of plant growth, especially humidity and light intensity. In any event, the roughly 200 million stomata on the leaves of a corn plant, each with a pore area of about 4×5 micrometers (1 micrometer is equal to a millionth of a meter) provide an open pore space of between 1 and 2 percent of the total leaf area. It is through this space that the movement of carbon dioxide, oxygen, water vapor, and other gases largely occurs.

In a normal day, stomata begin to open at or near sunrise, remain open during most of the day, and close in the late afternoon or early evening. Part of the opening and closing is regulated by the availability of water, part by light, and part by variation in the osmotic concentration, that is, dissolved salts and organic materials in the watery vacuoles, or central storage tanks, of the guard cells. In general, the osmotic concentration of guard cells is at least twice that of the surrounding, ordinary epidermal cells, but it goes up and down with the change in conditions around the plant.

The main osmotic component of guard cells is known to be potassium chloride. The content of this material tends to be high in the light and low in the dark, but it can also vary rhythmically and in response to plant hormones. One such hormone, abscisic acid (*see* "Turning Plants Off and On") is especially important. When the concentration of abscisic acid is high, potassium chloride leaves the guard cells. The resultant loss of osmotically active material causes a loss of turgor pressure, the closing of the stomata, and a diminution in the rate of transpiration. That these changes may be important in the physiology of the plant is shown by the fact that when a plant starts to wilt, its content of abscisic acid may rise tenfold in a few minutes. This rapid increase in abscisic acid in response to water stress may serve to shut down evaporative water loss in time to prevent excessive desiccation injury and death. The source of the extra abscisic acid, which appears so suddenly, is not yet known, but because of the swiftness of the change, it is believed to be released from some large molecule that holds it.

Paradoxically, partial closure of the stomata does not result in a proportional diminution in transpiration, since evaporation of water from spaced pores is proportional to the perimeter, not the area, of the pores, and perimeter does not change drastically with diminution of the pore from fully to half open. It is only when the pores are virtually closed that transpiration is measurably cut.

The action of light in causing the opening of stomata is intimately related to the concentration of carbon dioxide gas in the intercellular spaces just below the stomata. In normal air (.03 percent carbon dioxide) the guard cells are flaccid and the stomata are closed. When light hits the leaf and photosynthesis starts, the carbon dioxide concentration in these gas spaces is lowered, and the guard cells start to open. This goes on progressively until the carbon dioxide concentration is down to about .01 percent; further diminution in carbon dioxide results in no additional stomatal opening.

In extremely arid conditions, where excessive water loss readily leads to death, cultured plants must be irrigated, and wild plants must be well adapted to the environment to minimize water loss and maximize water uptake. Such adaptations include an extensive root system; a heavy, waxy leaf coat; and a diminished leaf area, so that fewer stomata are exposed to the evaporative forces of the habitat. Some xerophytes, or plants adapted to arid conditions, have no leaves at all, and their photosynthesis is restricted to the green and often fleshy stems. Those stomata that exist in these plants are sunk beneath the stem surface in depressions that minimize water evaporation through air movement.

In a world short of water and food, cultivation of arid zones might become more possible through the use of some technological control of transpiration. One thought is to cover plants with a plastic spray or sheet that would permit passage of carbon dioxide and oxygen but restrain or prevent passage of water vapor. The development of such a material would doubtless make the discoverer rich as well as renowned as a benefactor of mankind.

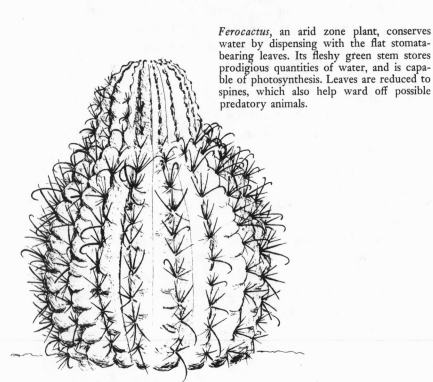

Ferocactus, an arid zone plant, conserves water by dispensing with the flat stomata-bearing leaves. Its fleshy green stem stores prodigious quantities of water, and is capable of photosynthesis. Leaves are reduced to spines, which also help ward off possible predatory animals.

4

The Membrane
Barrier

IN BIOLOGY, as in all fields of human endeavor, there are fads. For reasons that are not always clear, certain subjects, modes of thought, and approaches to problems predominate at certain times, thereby influencing the direction in which the field develops. Now that the initial excitement over DNA, RNA, and protein synthesis seems to have died down somewhat, problems associated with the cell membrane are the current "fad" in biology and have come to dominate much of the thinking of cell biologists. It is not that the membrane is a new discovery, but merely that new techniques of investigation and a new appreciation of the importance of spatial organization within cell systems have made this cellular component a central focus for much that is exciting in modern biology.

The excitement is certainly not generated by what one sees when looking at a membrane through even the most powerful microscope available. For the membranes that encase all cells—whether animal or plant—and all internal cellular units, or organelles, such as the nucleus, chloroplast, and mitochondrion, consist basically of only two layers of material separated by an apparently empty space. The layers always consist of proteins and fats, although these compounds differ in detail in different structures. The distance across the entire assembly is always uniform—about 10 nanometers, or one hundred-millionth of a meter. This is only about 1 percent of the cross-sectional diameter of even the smallest cells. Although small and inconspicuous, the membrane is one of the most important parts of the cell—without membrane integrity, no cell could long survive.

Certain cell membranes can be easily isolated in large quantities for chemical analysis. The best source of membranes to date has been the "ghosts" of mammalian red blood cells derived from rupturing these cells in water. When the cells burst, their contents spew out, and the envelope, which is practically pure membrane, can be removed from the other materials by high-speed centrifuging. Chemical analysis of the centrifuged membrane "pellet" reveals that it consists of about equal parts of two components—fats and proteins.

Most proteins are water soluble, while fats are oil soluble. Since oil and water do not mix, how can oil-soluble and water-soluble materials remain together in stable association in the membrane? The answer lies mainly in the special nature of the fats, or lipids, of the membrane. Like other lipids, they are built around a glycerol molecule and contain two chains of fat-soluble carbon atoms that are attached to one end of the glycerol. Also attached to the same glycerol molecule is a highly water-soluble portion of the lipid that terminates in a phosphate group. The resultant molecule, called a phospholipid, is a single, stable structure, one end of which is soluble in fat, the other in water.

The Membrane Barrier

This dual solubility furnishes the molecular basis for the stable, fat-permeable and water-permeable membrane. When pure phospholipids are dropped onto the surface of pure water, they form layers one molecule thick, with the water-soluble ends of the molecules sticking down into the liquid, and the fat-soluble carbon chains sticking up in the air. When the phospholipids are dropped onto oil, the reverse takes place. When they are dropped onto a medium composed of oil and water the phospholipid molecules partition themselves according to the solubility of each end, thus stabilizing the oil and water into an emulsion.

Proteins, which consist of long chains of amino acids, have both water-soluble, or hydrophilic, and water-rejecting, or hydrophobic, regions. When brought into contact with a phospholipid-stabilized emulsion, the protein molecule will twist and bend, orienting itself so that its hydrophilic parts are in the stabilized aqueous portion and its hydrophobic parts in the stabilized lipid portion of the emulsion. In this way, membranes can virtually assemble themselves into what appears to be a complex structure at the molecular level.

The membrane that encases all cells lies at the boundary between the cell and its external environment. Every molecule entering or leaving the cell must therefore deal with the barrier of the membrane. Just before the turn of the twentieth century, a physical chemist named Overton studied the penetration of various kinds of molecules into algal plant cells. He found that both fat-soluble and water-soluble materials enter the cell, but apparently through different pathways and by different mechanisms. Among fat-soluble compounds, the most soluble entered most rapidly; but among water-soluble compounds, the smallest molecules entered most rapidly. Overton concluded that the membrane must be a mosaic of fatty and aqueous pathways with the aqueous pathways being smaller than the fatty ones and thus tending to admit small molecules in preference to larger ones. Although Overton's mosaic idea has undergone many transformations and refinements, it remains essentially valid and is basic to the most modern models of the cell membrane.

A salient feature of this membrane is its differential permeability—the fact that certain kinds of molecules can penetrate it easily, others more slowly and with difficulty, and still others not at all. Because of this property, the membrane, situated at the doorway of the cell, determines, at least in part, which molecules can enter and which can leave the cell. But the membrane is much more complex than that: it has an inside and an outside, and both have an amazing capacity for change. To appreciate this, we must consider the different ways in which substances penetrate the membrane. In Overton's experiments, the penetration process seems to involve only simple diffusion. If there is a high concentration of a given substance outside the cell and a low concentration inside the cell and the membrane contains a pathway for this compound, the substance will tend to enter the cell. Conversely, if there is a high concentration of the substance inside the cell and a low concentration outside, the compound will tend to leave the cell.

For many water-soluble substances, entry into the cell first involves actual attachment to a specific site on the membrane, probably the location of a specific protein. A simple experiment will illustrate. Let us take a cell, immerse it in a solution of a simple salt such as potassium chloride, and measure the rate at which potassium enters the cell. If we now add to the external potassium chloride solution a closely related salt, rubidium chloride, the potassium will enter the cell less readily than previously. By varying the ratio of potassium to rubidium in the external medium and measuring the rate of penetration of each, we can prove that the two salts are competing for space at the same membrane pathway. It has been found that sodium chloride, another closely related salt, enters the cell by yet another pathway, at which other substances in turn compete with it for entry. From experiments like this, we can conclude that the aqueous part of the mosaic of the cell membrane has specific transport sites for the entry of particular materials; each site is presumed to owe its specificity to the presence of a single protein or group of proteins.

The penetration of some substances, however, is not a matter of passive diffusion and entry in accordance with the concentration, but may involve the performance of work by the cell. For example, in order to pass through certain membranes, ordinary sugar, or sucrose, must first be transformed chemically by the attachment of a phosphate group. The sugar traverses the membrane in the form of sucrose phosphate, and once it has entered the cell it is reconverted to sucrose by a reaction that splits off the phosphate group. The phosphate is therefore a carrier for the entry of sucrose into the cell, but the work of adding and eliminating the phosphate group involves energy-using chemical reactions. Because of the external location of the phosphate-attaching reaction and the internal location of the phosphate-removing reaction, sucrose transport through the membrane can proceed in only one direction—outside to inside. Hence, such a membrane is said to have "directionality."

The permeability of a membrane to a given material is not a fixed property but may change in response to a number of physical and chemical stimuli. Hormones, for example, frequently cause the cell to alter its uptake or release of a particular dissolved substance, or solute. In addition, the solute itself can induce the cell to change its permeability characteristics. Usually, the cell responds to the latter stimulus by making new proteins in the membrane, which combine with the solute in question to facilitate its entry. Through this means, a cell can handle one substance at a time. Given two substances in a mixture, certain microorganisms will first take up all of one of them before absorbing any of the other. One substance evidently prevents absorption of the other by inhibiting formation of the specific protein needed for its uptake. Membranes thus have a dynamic quality and are constantly changing the nature and location of their absorption sites.

The mitochondria and chloroplasts are cellular substructures that deal with the conversion of energy. In the mitochondria, found in all cells, fuels such as sugar and organic acids are oxidized. The energy released in the process is captured in the special phosphate bonds of energy-rich compounds like adenosine triphosphate, or ATP, which is manufactured by the mitochondria. It is ATP that, among its other activities, furnishes the phosphate group that moves sucrose across the cell's surface membrane as sucrose phosphate. ATP is therefore a kind of "energy currency" that is spent when cellular work needs to be done. ATP is also made in the chloroplast, the organelle in plant cells in which photosynthesis takes place. Highly structured specific membranes are involved in the chemical reactions that take place in both the mitochondria and chloroplasts. The resultant flow of elementary particles—electrons and protons—through particular protein pathways in the organelle membranes can cause the formation of ATP. But if the membrane structure is disrupted by mechanical or chemical means, ATP formation no longer occurs, even in the presence of all the necessary components. The integrity of membrane structure is therefore crucial.

The Membrane Barrier

Membranes have recently acquired a new importance in cancer research. If ordinary animal cells are placed in a chemical medium favorable to their growth, they will divide and grow until they cover the entire surface with a single layer of cells. Then, due to a mysterious process called "contact inhibition," they stop growing. When cancerous cells are placed in a similar medium, they show no contact inhibition but grow and pile up on top of one another without stopping. This leads to the theory that surface membrane proteins are involved in the normal cessation of growth and that cancer cells may have abnormal surface proteins. If this theory is correct, then immunological approaches to cancer would seem appropriate.

The surface membranes of both normal and cancerous cells are in a state of constant activity, which is sometimes visible by time-lapse photography under a microscope. They are able to send out projections that can surround and engulf ambient particles. In the case of the single-celled amoeba, the particles may be food; with mammalian cells, the particles may be invading bacteria being combatted by a white blood cell. When the captured particle is brought back to the surface of the ingesting cell, the membrane pinches together, cutting off the projection, or vesicle, surrounding the foreign particle. The vesicle then moves to the interior of the cell, where it can be digested by cellular enzymes. It is apparent that although the cell membrane may exclude many small molecules through differential permeability, it can ingest relatively large particles. Even large pieces of DNA can enter cells by this process, leading to the possibility of direct genetic transfer from one cell to another.

Certain cells, like the rods and cones in the retina of the eye and the nerve cells that transmit electrical stimuli, have special membranes designed for transforming chemical reactions into electrical signals. Nerve cells, for example, show electrical changes because the membrane periodically changes its electrical charge and releases large quantities of salts through diffusion channels. The channels then close up, and the salt accumulation in the nerve cell is restored by a process that involves a so-called pump, or protein, that uses ATP energy to move one salt across the membrane. Such pumps are present in all cells, and research into how they work is one of the most fascinating aspects of the current inquiry into the cell.

5

The Blind Staggers

IN THE SUMMERS of 1907 and 1908, about 15,000 sheep died suddenly and unexpectedly in a specific area north of Medicine Bow, Wyoming. The Wyoming State Board of Sheep Commissioners attributed the deaths to acute poisoning by toxic plants, generally considered to be members of the vetch group (genus *Astragalus*) and the woody aster group (genus *Xylorrhiza*). Within the next several decades similar occurrences were reported throughout a region that included Colorado, Wyoming, Utah, South Dakota, Nebraska, Kansas, and parts of Montana. In each case, the stricken animals, usually sheep but sometimes cattle and occasionally horses, developed symptoms of either alkali disease or "the blind staggers" after grazing heavily in certain areas.

Of the two diseases, the blind staggers is far the more severe. Animals that have eaten *Astragalus* or *Xylorrhiza* start to wander in circles, stumbling over any obstacles in their way. Their appetites and thirst diminish; then their posture alters, their front legs weaken, and their vision becomes impaired. Still later, symptoms of paralysis appear, tongue movements and swallowing reflexes are blocked,

Astragalus Xylorrhiza

Astragalus and *Xylorrhiza*. These two plants accumulate large quantities of selenium in the form of amino acids, in which the selenium substitutes for sulfur. Such material is highly toxic to animals that consume these plants.

severe abdominal pains follow, and ultimately, respiratory failure leads to death.

The less dramatic symptoms of alkali disease are loss of hair, especially from the tail, and the cracking and partial sloughing off of the hooves. Hoof material regenerates periodically, but the cracking causes such extreme tenderness that animals suffering from the disease are unable to graze well. It has proved impossible to raise to marketable size livestock affected with alkali disease, and many ranchers have gone bankrupt in the attempt.

As livestock losses from these two diseases increased over the decades with no known means of prevention, the ranching industry turned to science to unravel the mystery. If the suspected plants did indeed contain a material sufficiently poisonous to cause alkali disease and the blind staggers, why did their toxic action occur only in certain restricted areas? The solution turned out to be as bizarre as any in toxicology, for the chief clue was selenium, an element almost as rare as gold. Discovered in 1817 by the great Swedish chemist Jöns Jakob Berzelius, selenium is generally found in soil at a concentration of only about 2 parts per million, although in rare cases the concentration has been known to go as high as 100 parts per million.

Selenium is thought to be of igneous, or volcanic, origin; it occurs usually in the form of salts such as iron selenide in Cretaceous rocks, outcroppings of which are scattered throughout the rangelands of the western United States. The relevant point for the saga of the blind staggers, however, is that both *Astragalus* and *Xylorrhiza* possess a prodigious ability to absorb from soil large quantities of compounds containing selenium. *Astragalus bisulcatus*, the most poisonous of the leguminous plants, is able to accumulate up to 5,530 parts per million of selenium from soils with concentrations of only about 2 to 100 parts per million.

Xylorrhiza and other toxic species accumulate selenium somewhat less powerfully, but all plants that cause the blind staggers have this same remarkable property.

When *Astragalus bisulcatus* plants are grown in a nutrient solution in which the amounts of the various mineral elements are carefully regulated, the selenium content produces some unexpected effects. So far as is known, selenium is not absolutely essential to the growth of any plant, but it clearly enhances the growth of *Astragalus bisulcatus* and related plants. The dry weight and height of an *Astragalus* fed modest amounts of selenium may be five to ten times greater than one totally deprived of the element. This probably accounts for the unusual adaptation of *Astragalus* plants to an ecological niche that is rich in seleniferous rocks. These plants act as wicks, drawing in the element from the surrounding rock and causing a high concentration to accumulate in the area around their roots. The selenium content of the soil around such plants could theoretically become high enough to cause direct injury to nearby plants of other types. This has not been definitely proved in nature, but laboratory experiments have shown that plants unable to accumulate selenium may be dwarfed and show leaf bleaching when subjected to the same levels of applied sodium selenite, a usual source of the element that stimulates growth in *Astragalus*. The roots of the inhibited plants frequently thicken and develop suppressed lateral roots of a pinkish color due to intracellular deposits of elemental selenium.

Why does the selenium absorbed by a plant cause extreme toxicity upon its ingestion by an animal? First, it should be noted that, unit for unit, selenium in an *Astragalus* plant is much more poisonous for an animal than is the inorganic salt, sodium selenite. This means that the plant must transform accumulated sodium selenite into some other, more toxic form. Furthermore, it has been shown that all plants collect selenium far more efficiently if it is fed to them as an extract of *Astragalus*, rather than as a selenite salt. Chemical fractionation in the laboratory of the water-soluble selenium-containing extract of *Astragalus* reveals an unexpected aspect of the plant's metabolism.

It is well known that selenium is a chemical analog of sulfur. This element, essential to all living cells, has a variety of functions in the organism, the most important of which is its presence in almost all proteins. Proteins are composed of some twenty different amino acids linked in a specific order, and two of the twenty, cysteine (or cystine, its oxidized form) and methionine, contain sulfur. It appears that selenium imitates sulfur so closely that the *Astragalus* plant has come to substitute it for sulfur in its protein chain by making analogs of the two sulfur-containing amino acids, which are called seleno-cysteine and seleno-methionine. And when ingested by animals, these compounds produce disastrous symptoms of toxicity. Their toxicity is probably related to their competition with the normal sulfur-containing amino acids for incorporation into proteins. Thus, animals that receive amino acids containing selenium in effect suffer from an insufficiency of the amino acids that contain the essential sulfur.

When, after grazing mainly on grass that contains about ten to thirty parts per million of selenium, animals develop symptoms of alkali disease, they may be cured by changing to a diet lacking selenium. Supplemental feeding of proteins rich in sulfur-containing amino acids also can prevent the development and help alleviate the toxicity symptoms of the selenium-containing diet. But when animals ingest *Astragalus* with more than 5,000 parts per million of selenium, nothing can be done; they are doomed to develop the blind staggers disease and die.

There are also other interesting aspects to the problem

of selenium toxicity in the field. Plants containing selenium, like their sulfur-containing analogs, frequently have an offensive garlicky odor. Horses are choosy eaters and tend to avoid such plants; thus they are rarely afflicted with the disease unless other plants are unavailable. Cattle are much less selective and tend to graze indiscriminately, thus occasionally consuming large quantities of *Astragalus* and especially *Xylorrhiza*, which grows abundantly on the desert floor. Sheep actually seem to prefer *Astragalus* and *Xylorrhiza* to other foods and seek them out. This difference in nutritional habits accounts for sheep being the most common victims of the blind staggers.

About half a pound of *Astragalus bisulcatus* will kill a sheep thirty minutes after ingestion; smaller quantities take a longer time to produce their effects. About 400 to 800 parts per million of selenium are fatal when ingested at the rate of about 8 to 16 grams per kilogram of body weight (about ¼ ounce per pound of body weight). After an animal ingests *Astragalus* material, selenium is found in all organs, but it appears to be most abundant in the liver, blood, kidneys, spleen, and brain. It is eliminated from the body mainly by means of the urine; some is also lost in the feces, in the sweat, and in the breath.

A curious sidelight to this story rests in the properties of another species of *Astragalus*, called locoweed. This plant, too, causes similar symptoms and poisoning by virtue of an organic molecule, but it does not contain high quantities of selenium, and so far as is known the effective mechanism is not related in any way to selenium metabolism.

What can be done to combat these devastating diseases, produced not by a grazing plant itself, but by the plant's absorption of an element that in itself would not normally be toxic? The obvious tactic would be to prevent grazing in the regions where *Astragalus*, *Xylorrhiza*, or other seleniferous plants thrive. But this would rule out vast areas of the western range at a time when the supply of food is becoming more and more critical. Another possible tactic would be to treat the area with herbicidal chemicals that would kill the offending plants. But since these plants tend to dominate seleniferous areas because their growth is promoted by selenium, tremendous quantities of herbicides would be required. Such an operation would therefore seem economically unfeasible and also ecologically unattractive, as no herbicide is without some undesirable side effects. A third possibility would be to swamp the area with sulfur-containing materials, which the plants would absorb. The ability of *Astragalus* or other seleniferous plants to convert inorganic selenium to seleno-amino acids depends on the ratio of sulfur to selenium. Even when this ratio is about three to one, optimal incorporation of selenium into the toxic amino acids occurs, but if the proportion of sulfur were raised, the formation of the seleno-amino acids would tend to decline. Although heavy sulfate fertilization is theoretically a possible way of mitigating the trouble, it is unlikely to become a practical, effective weapon against selenium poisoning. At the moment, however, while there seems to be no permanent resolution to the problem, the blind staggers is not a serious menace because ranchers have learned to avoid seleniferous areas for grazing.

What danger does this disease pose to humans? Practically none, it would seem. Presumably if humans were to ingest amino acids containing selenium, they, too, would develop symptoms equivalent to alkali disease or the blind staggers. But, in fact, the afflicted animals show such terrible symptoms that they are always identified and never used for meat. Consumption of grains grown in the seleniferous regions is also avoided. Nonetheless, since selenium can be found everywhere in small quantities, we may well ask whether enough selenium is present in ordinary diets, either as inorganic selenides or as organic compounds, to affect human health—a question that public health authorities should investigate more thoroughly.

The Plant
Coordinates Itself

6

Botanist

Charles Darwin

HAD CHARLES DARWIN never contributed to the theory of organic evolution, he would still have gone down in history as a remarkably gifted and intuitive experimental botanist. Many plant physiological investigations of today are based on leads that he uncovered. Entire areas, such as studies of hormonally controlled correlations, can be said to have developed directly from his simple experiments and the observations he made.

Born in 1809, Darwin was the grandson of Erasmus Darwin, a well-known zoologist and botanist, and a cousin of Sir Francis Galton, the anthropologist who also contributed so fruitfully to biostatistics and eugenics. Nothing about his early life foreshadowed his noteworthy future. He was a wealthy, rather indolent, and physically weak young man whose early educational ventures were disasters. Of the Shrewsbury School, where he studied under Samuel Butler, he says in his autobiography, "The school as a means of education to me was simply a blank." * He did not know what to do with himself after graduation and reports his father as exclaiming in mortification, "You care for nothing but shooting, dogs, and rat-catching, and you will be a disgrace to yourself and all your family." Were such a young man to apply to one of today's better universities, he would undoubtedly be denied admission.

In 1825, at the age of 16, he was sent by his father to Edinburgh University, where Darwin's brother was studying medicine, with the intention that Charles also enter medical training. As with many parental plans for children, this one didn't take. Charles himself explains, "I became convinced from various small circumstances that my father would leave me property enough to subsist on with some comfort, though I never imagined that I should be so rich a man as I am; but my belief was sufficient to check any strenuous effort to learn medicine."

University courses in zoology and geology repelled him. He says of them, "The sole effect they produced on me was the determination never as long as I lived to read a book on geology, or in any way to study the science."

Nevertheless, it was at Edinburgh that Darwin's interest in natural science was aroused. While attending the university he took part in the activities of the Plinian Society, which met to read papers on the subject.

In 1826, Darwin transferred to Cambridge University to study theology, again with disastrous results. "During the three years I spent at Cambridge my time was wasted, as far as the academical studies were concerned, as completely as at Edinburgh and at school," he writes in his autobiography. "I attempted mathematics, and even went during

*All the quotations in this chapter are taken from *The Darwin Reader*, edited and annotated by Marston Bates and Philip S. Humphrey (New York: Charles Scribner's Sons, 1956).

Botanist Charles Darwin

the summer of 1828 with a private tutor (a very dull man) to Barmouth, but I got on very slowly." Fortunately, at Cambridge Darwin became acquainted with a professor of botany, J. S. Henslow, whom he accompanied on botanical walks. Darwin ultimately became a friend of the Henslow family, frequently sharing dinners and scientific conversation with their guests. It was through Henslow's friend Captain Fitz-Roy that Darwin was invited to join the crew of the *Beagle* as the official naturalist. The cruise of the *Beagle* changed his entire life, laying the foundation for all his subsequent scientific interests.

From 1839, when he published his research from the *Beagle*, to 1859, when the *Origin of Species* appeared, Darwin's main concern was evolutionary theory. Geologic, zoological, and botanical data were of course interwoven in the base from which the theory arose. His first mainly botanical publication was *The Various Contrivances by Which Orchids are Fertilized by Insects*, published in 1862. The data for this work had been collected for more than twenty years, as the autobiography recounts.

During the summer of 1839 I was led to attend to the cross-fertilization of flowers by the aid of insects, from having come to the conclusion in my speculations on the origin of species, that crossing played an important part in keeping specific forms constant. I attended to the subject more or less during every subsequent summer.

This work led Darwin to a study of dimorphic flowers in primroses (*Primula*) and other species. The significance of dimorphism in plants is that it protects against self-pollination and thus permits greater variation in evolution.

I do not think anything in my scientific life has given me so much satisfaction as making out the meaning of the structure of these plants [*Primula*]. I had noticed in 1838 or 1839 the dimorphism of *Linum flavum* [flax], and had at first thought that it was merely a case of unmeaning variability. But on examining the common species of *Primula* I found that the two forms were much too regular and constant to be thus viewed. I therefore became almost convinced that the common cowslip and primrose were on the highroad to become dioecious;—that the short pistils in the one form, and the short stamens in the other form were tending towards abortion. The plants were therefore subjected under this point of view to trial; but as soon as the flowers with short pistils were fertilized with pollen from the short stamens, they were found to yield more seeds than any other of the four possible unions, and the abortion-theory was knocked on the head. After some additional experiments, it became evident that the two forms, though both were perfect hermaphrodites, bore almost the same relation to one another as do the two sexes of an ordinary animal. With *Lythrum* [loosestrife] we have the still more wonderful case of three forms standing in a similar relation to one another. I afterwards found that the offspring from the union of two plants belonging to the same forms presented a close and curious analogy with hybrids from the union of two distinct species.

Some of Darwin's other investigations may yet be the precursors of new pathways in experimental botany. About 1860, he became interested in insectivorous plants, especially the sundew (*Drosera*), whose sticky tentacles move to surround and digest any insect that chances to land on them. Darwin found that sundews can digest meat and other proteins by means of excreted enzymes. As a result of feeding experiments, he found that it was not only proteins that caused the plant's reactions but that simple salts were also effective. He listed all the salts he tested according to their activity, and discussed the effects of acids, alkaloids, and many other groups of substances. Each of his experiments was carried out with the proper controls, and in cases where no response was obtained he often treated the leaves with ammonium carbonate to ascertain whether they were still sensitive or not to any stimulus. Even a statistician would be satisfied with Darwin's experiments since in all his work he used many thousands of leaves.

In 1858, he read a short paper on tendril movements in wild cucumber plants written by Asa Gray, the famous American botanist. The subject fascinated him; he obtained seeds from Gray, raised the plants, and became enmeshed in analyzing the revolving movements of tendrils and stems. Through a period of ill health, he continued these investigations and in 1864, sent a long paper on climbing plants to the Linnean Society, a prestigious British botanic group founded in the eighteenth century. In 1875, the paper was revised and published as a separate book, *The Movements and Habits of Climbing Plants*. Then in 1880, aided by his son Francis, who had worked for a year in Germany with the famous plant physiologist Julius Sachs, Darwin wrote *The Power of Movement in Plants*, a book that has considerably influenced modern plant physiology.

Darwin carefully observed the movements of growing plants by devising some homemade laboratory equipment. He would attach a slender glass filament, whose end had been dipped in a bright pigment, to a plant tip to amplify its movements. Then, placing a glass plate over the plant, he made dots on the glass to mark each new tip position. By connecting the dots, he obtained a continuous recording of the plant's movements, which he called circumnutation.

After studying the movement of plants in the absence of any external stimulus, Darwin initiated a detailed study of the effects of gravity and light on plant movements. Exposing young seedlings of the grass family to unilateral

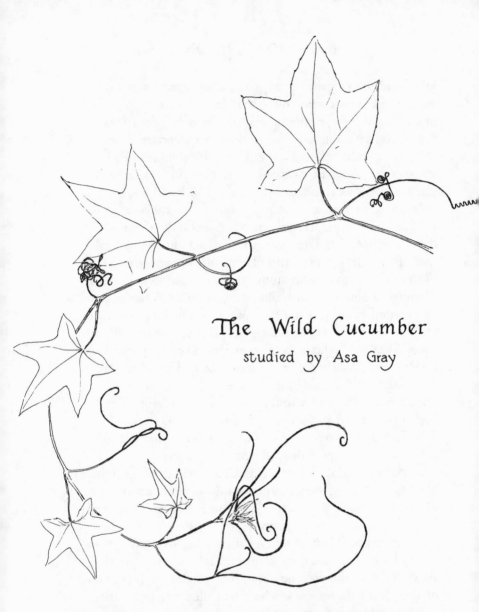

The Wild Cucumber
studied by Asa Gray

The wild cucumber, studied by the distinguished nineteenth-century American botanist, Asa Gray. The tendrils of this plant are extraordinarily sensitive to touch. Once in contact with a rigid body, their previously straight growth pattern is altered into a vigorous coil, which fastens the plant body to the support. This device permits the plant to grow to great heights and over great areas without having to make much supporting woody tissue in its stems.

light, he noted an alteration of the usual circumnutational pattern and a bending of the plant toward the light (phototropism). By covering the seedling with a minutely scored, dark-lacquered glass shield through which pinholes of light were transmitted, Darwin ascertained that only the extreme tip of the grass could perceive the light energy, while cells that were some distance from the tip were responsible for the actual curvature. He therefore postulated the existence of some mechanism for the transmission of the light stimulus from one plant part to another.

Concerning the response of plants, especially their roots, to gravity, a reaction referred to as geotropism, Darwin noted that removal of the tip would annul further response.

The fact of the tip alone being sensitive to the attraction of gravity has an important bearing on the theory of geotropism. Authors seem generally to look at the bending of a radicle towards the center of the earth as the direct result of gravitation, which is believed to modify the growth of the upper or lower surfaces in such a manner as to induce curvature in the proper direction. But we now know that it is the tip alone which is acted on, and that this part transmits some influence to the adjoining parts, causing them to curve downwards. Gravity does not appear to act in a more direct manner on a radicle, than it does on any lowly organized animal, which moves away when it feels some weight or pressure. . . . It is hardly an exaggeration to say that the tip of the radicle thus endowed, and having the power of directing the movements of the adjoining parts, acts like the brain of one of the lower animals; the brain being seated within the anterior end of the body, receiving impressions from the sense organs, and directing the several movements.

These statements created a storm among plant physiologists, especially in Germany. Most investigators thought it necessary to try to disprove Darwin's observations, and many papers were published to show that the tip was no more sensitive to light and to gravity than the rest of the growing stem. However, thirty years later, Peter Boysen-Jensen in Copenhagen initiated experiments on localized phototropic perception that confirmed Darwin's hypothesis. The tip did indeed control the growth of the subjacent region through some influence that moved downward. This influence, presumably chemical, could pass through a wound and even through an aqueous gelatin barrier, but was stopped by a mica or metal barrier. Its distribution was symmetrical in unstimulated plants, but asymmetrical in unilaterally light-stimulated plants. These findings led in 1928 to the discovery by Frits Went in Utrecht of *auxin*, the first defined, mobile chemical stimulus in plants. Went made this discovery while still a graduate student.

When the leaf sheath of a grass seedling is exposed to unilateral light (right), the growth hormone *auxin* is diverted to the shaded side (heavy arrows). In the dark (left) the flow of the hormone is symmetrical on the two sides of the organ.

The fact that the extracted and chemically characterized *auxin* satisfactorily explained the influence of the tip on subjacent regions established plant hormone studies as an important part of experimental botany. This field has continued to grow and now dominates modern agriculture. Virtually no agricultural product is produced today without some help from a plant hormone or a synthetic analog. Such substances control almost all phases of plant life, including seed germination, stem growth, root initiation, bud

Anyone who has grown plants in a north-facing window knows they turn desperately towards the light.

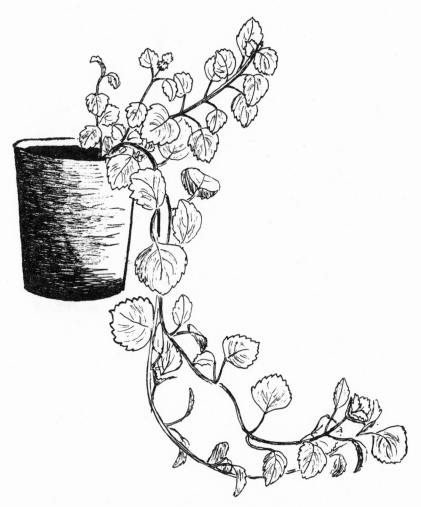

A plant of Swedish ivy showing several tropistic responses. Stems grow upward, away from the center of the earth, but also respond to uni-lateral light, in this instance coming from the right. Even the prostrate stem, whose weight has caused it to fall below the surface from which the plant is growing, is struggling to curve upward at its tip.

and seed dormancy, flower formation, fruit set, fruit ripening, senescence, and leaf fall, or abscission.

Darwin's work, carried out only to satisfy his consuming interest in plants, has thus had profound practical applications for agriculture. Had he been working for some chemical company and been instructed to produce a herbicide, a fruit ripener, an abscission or antiabscission agent, he might or might not have been able to do so. But his fertile imagination, sharp powers of observation, and experimental virtuosity led him to many important discoveries.

It is characteristic of Darwin that he was not satisfied with his own accomplishments. He says of himself:

> My mind seems to have become a kind of machine for grinding general laws out of large collections of facts, but why this should have caused the atrophy of that part of the brain alone, on which the higher tastes depend, I cannot conceive. A man with a mind more highly organized or better constituted than mine, would not, I suppose, have thus suffered; and if I had to live my life again, I would have made a rule to read some poetry and listen to some music at least once every week; for perhaps the parts of my brain now atrophied would thus have been kept active through use. The loss of these tastes is a loss of happiness, and may possibly be injurious to the intellect, and more probably to the moral character, by enfeebling the emotional part of our nature.

No better plea for a liberal education has ever been made.

7

Which End Is Up?

WHEN a farmer or a gardener throws a seed into the ground, he doesn't have to worry which end of the seed is up and which down. Whatever the orientation happens to be, and barring the unusual, the root will go down and the stem will come up. From a functional point of view, this is, of course, as it should be. The root is designed to anchor the plant and to absorb water and mineral nutrients for growth and development. To do these things, it must penetrate downward into the soil. The stem, on the other hand, is an axis for holding the maximum number of leaves in the best position to carry out the process of photosynthesis through light absorption. Obviously this activity has to be carried on above ground where the light is. But how does the plant distinguish up from down? What mechanism determines that the root will go down and the stem go up? These are complicated questions to which we have only partial answers despite a century of investigation.

If a germinated seedling is removed from the medium in which it has been growing and is laid on its side, the root will turn down, toward the center of the earth, and the stem will bend upward, away from the center of the earth.

This curvature response is called geotropism. It is brought about by unequal rates of growth on the two sides of the cylindrical plant axis. In a horizontally placed seedling, the upper side of the root grows more than the lower side, and the resulting differential growth points the root tip down. The situation is reversed with the stem. When placed horizontally, the lower side of that organ grows more rapidly than the upper side, and the resultant growth curvature is therefore upward.

If an erect plant (left) is laid on its side (right), the stem will turn upward and the roots downward.

Seeds germinated in a satellite circling high above the earth, where the planet's gravitational field is very low, do not orient with their roots pointing down toward the center of the earth. This indicates that at least part of the reaction of seeds at the earth's surface is due to gravity. The gravitational stimulus, however, can be overcome by centrifugal force. Thus, if a plant is tied to the outer edge of a turntable that is spun with a force equivalent in the horizontal plane to the downward pull of gravity, the stem will grow partly inward instead of straight up, and the root partly outward instead of straight down. The root will, in effect, respond to a simple vectorial problem in which equal forces are exerted parallel and perpendicular to the earth's surface, and grow at an angle 45° below the horizontal. As the centrifugal force of the rotating turntable is progressively increased, the angle of the root will approach the horizontal until the root is parallel to the turntable.

These facts suggest that the gravitational response of roots and stems might be due to the dislocation of a heavy body, or statolith, within plant cells, which can also be displaced by centrifuging. This is called the statolith theory. It presumes a mechanism similar to that by which animals balance themselves through the displacement of small calcium carbonate crystals in the semicircular canals of the ear. What could serve as a statolith in a plant cell? It appears that for the purpose of up–down orientation, plants employ very dense starch grains that fall to the bottom of specialized cells in response to gravity. If a root or stem is tipped 90°, the heavy starch granules will roll from the bottom to another wall of the cell, thus indicating a new direction for the down–up axis. This theory is supported not only by direct microscopic observation of such falling starch grains but also by experiments in which removal of the granules from the specialized cells of the root cap bearing such starch grains makes the root insensitive to gravity. If, for example, the cap is carefully removed from certain roots, the rate at which these roots grow is not affected, but the organs' ability to respond to gravity is lost and the roots will grow in whatever orientation they are placed. When the roots regenerate new caps containing visible statoliths, they regain their ability to curve geotropically. Although we cannot always see statoliths in geotropically sensitive cells, we reason by analogy that what is true in one system probably holds true for many systems.

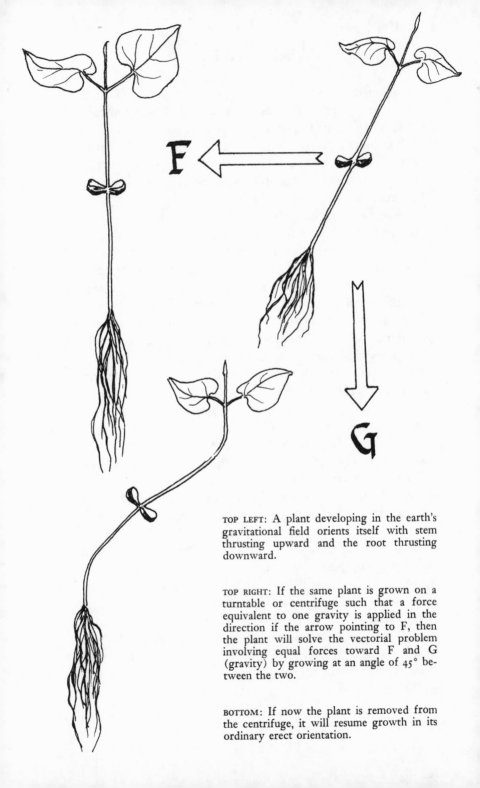

TOP LEFT: A plant developing in the earth's gravitational field orients itself with stem thrusting upward and the root thrusting downward.

TOP RIGHT: If the same plant is grown on a turntable or centrifuge such that a force equivalent to one gravity is applied in the direction if the arrow pointing to F, then the plant will solve the vectorial problem involving equal forces toward F and G (gravity) by growing at an angle of 45° between the two.

BOTTOM: If now the plant is removed from the centrifuge, it will resume growth in its ordinary erect orientation.

Which End Is Up?

How does the rubbing of the statolith against the cell wall create a disturbance in growth that results in geotropic curvature? The reaction is probably connected with an effect on the surface membrane of the cell's protoplasm, which presses against the cell wall. This effect, in turn, leads somehow to an altered transport of growth-promoting and growth-inhibiting substances. For example, a vertically oriented stem tip or root tip contains roughly equal quantities of growth-promoting and inhibitory hormones in its two lateral halves. But if a root tip is laid on its side, more growth hormones and growth inhibitors will be obtained from the lower than the upper side of the displaced tip. Since root growth is inhibited by the new, higher concentrations of hormone and inhibitor, we reason that the curvature is due to an inhibition of growth of the root's lower half relative to its upper half. With stems that have been horizontally placed, one also finds an accumulation of growth hormone on the lower side. But here, the higher concentration of hormone leads to a higher growth rate and thus to upward curvature of the stem.

Asymmetry in the distribution of growth-regulating substances is correlated with the development of a transverse electrical potential of up to 100 millivolts in a horizontally displaced organ. It used to be thought that this electric potential developed soon after statolith displacement and was the cause of the flow of negatively charged growth hormone toward the positively charged side of roots and stems. But recent experiments have shown that the reverse sequence of events is true; the negatively charged growth hormone is first displaced asymmetrically and then the transverse electrical potential arises. Undoubtedly the two phenomena are connected in an important, but unknown, way.

Putting everything together, our theory now reads as

follows: Starch grains are displaced, stimulating the cell membrane; asymmetric distribution of growth regulatory substances and a transverse electrical potential follow; this leads to asymmetric growth and curvature.

While these simple facts provide an explanation of sorts about the positive geotropism of roots (toward the force of gravity) and the negative geotropism of stems (away from the force of gravity), they still leave many questions unanswered. For example, why does a stem grow in one direction and a root in another? The glib answer is that the growth regulator displacement leads to inhibition of root growth and promotion of stem growth. But quantitative investigations of the amounts of hormone displaced compared with the amount required to produce enough growth differences to lead to curvature tell us that this theory is probably not too accurate. That is why growth inhibitors are now surveyed as well. A second and more perplexing question is this: How can the rubbing of a starch grain against the cell membrane lead to lateral displacement of hormone and the appearance of bioelectric potentials? This may have to do with the movement of ions of hydrogen or a salt, since an electrical charge arises when a freely diffusible ion accumulates on one side of a differentially permeable membrane. But how can the rubbing of a starch grain lead to the movement of many millions of ions? The answer to this question remains unclear. Much work thus must still be done on the physiology of geotropism.

Not only stems and roots but also the petioles, or stalks, of leaves are sensitive to gravity and orient the leaf blade so it is roughly parallel to the earth's surface, and thus roughly perpendicular to the incoming solar radiation. When a potted plant is put on its side, its leaves reorient. This response, too, presumably involves the gravity perception of statoliths.

Which End Is Up?

Even large tree trunks displaced from the vertical can right themselves in a most remarkable way by the formation of what is called "reaction wood." When a conifer stem is placed horizontally, the cell division activity of the cambium, the thin layer between wood and bark that produces new wood cells, becomes asymmetric; cells on the lower side of the stem will be much thicker walled and impregnated with much more of the woody substance lignin than cells on the upper side. This effect leads ultimately to a tremendous strain within the stem, which bends the trunk mechanically in a direction that tends to right it. Because they tend to press the stem upward, the heavy-walled cells on the lower part of the angled trunks of conifers are referred to as compression wood. Hardwood trees, on the other hand, give rise to a different kind of special cell

A pine tree which has grown in the erect orientation has symmetrical annual rings.

A pine tree which has been prostrated produces more wood on its lower side than on its upper side. This tends to bend the trunk upward toward the erect position.

on the upper part of a prostrate trunk. These cells are especially low in lignin and act as tension wood, in effect producing a gradient that again causes the rigid tree trunk to curve upward. Whether starch grains, hormones, and growth inhibitors are involved in this process, which can generate tremendous forces, is not really understood.

Some of the new growth-regulatory chemicals that are used extensively in agriculture interfere with the transport of growth hormones in a stem. Thus, the hormone auxin normally moves only basipetally, that is, from the stem tip down to the root tip, and many plant growth reactions are the result of this unidirectional movement. But the application of a synthetic growth regulator such as triiodobenzoic acid, which is used to increase soybean yield, can interrupt the flow of auxin and cause it to accumulate in unexpected places. When this happens, tropistic responses are upset; some roots may be observed growing upward within the soil, and stems may curve and become parallel to the earth's surface or may actually turn downward. These abnormalities underscore the delicacy of the regulatory system of plants and emphasize the need for care in the use of such chemicals.

Plants can determine which end is up and which down, at least as long as we don't flood them with noxious chemicals or spin them too dizzily on a rotating wheel. But such treatment would probably cause humans to lose their sense of orientation too.

8

Rotten Apples and Ripe Bananas

THERE IS TRUTH in the folk saying that one rotten apple will spoil a whole barrelful of fruit. In recent years, it has been shown that plant tissue either mechanically injured or invaded by a fungus produces large quantities of a simple, well-known hydrocarbon gas—ethylene. This hydrocarbon occurs in the natural gas widely used as heating and cooking fuel. When supplied to healthy plant tissue, the gas produces some remarkable and drastic consequences. One of these is to stimulate the healthy tissue also to produce ethylene. Thus, one whiff of ethylene produces a cascade effect, and soon the entire barrel is full of ethylene-producing fruits. When the concentration of ethylene gets sufficiently high, the fruit tissue goes through the transformations we normally associate with ripening and rotting—softening, sweetening, darkening, and eventually, dissolution.

The adverse effects of ethylene on plant growth and development have been known for many years. In the

Even one rotten apple in a barrel produces so much *ethylene*, the plant's ripening hormone, that it makes the other apples ripen. They, in turn, produce more ethylene, ripening yet other fruits. The ripened fruit then produces characteristic odors, which attracts fruit flies and other insects.

1920s an American grower of chrysanthemums had a greenhouse full of plants being readied for the market, when the weather turned cold. He decided to heat his greenhouse with a burner system operating on natural gas, a method he had previously used without mishap. This time, unfortunately, the burner system was not properly adjusted, and some of the uncombusted gas was released into the greenhouse. When the grower returned, he found that all of the plants had dropped their leaves. His crop was completely unmarketable. The economic loss was great enough to spur research into the identification of the toxic agent. After a few years, it became clear that unsaturated hydrocarbons were responsible: common gases like ethylene, propylene, butylene, and acetylene were found to be effective defoliators, and ethylene was as active as any other compound. To protect plants from the deleterious effects of natural gas, it was sufficient to trap the ethylene and related gases by simple chemical techniques, such as bubbling the gas mixture into bromine water. That process works because two atoms of bromine add onto the ethylene molecule, converting it into nonvolatile ethylene dibromide. Saturated hydrocarbons, such as ethane, propane, and butane do not undergo such an addition reaction with bromine and are accordingly swept away with the effluent gas stream. Gas deprived in this manner of its unsaturated components produces no harmful effects on plants.

What can be done with a barrel of apples that constantly produce their own ethylene? Continually sifting the barrel contents to remove an apple the moment it begins to look rotten is obviously impractical. It is also not economically practical to bubble the air in the barrel through bromine water at frequent intervals. A much simpler method is to add a compound that will counteract the effect of ethylene on the healthy apple tissue. Such a compound is another simple gas, carbon dioxide, produced by all living cells in the process of respiration. The normal concentration of carbon dioxide in the air is 0.03 percent of the total. At that low level, carbon dioxide is unable to counteract the effects of even small quantities of ethylene. But when the carbon dioxide concentration of a closed container is raised to 5 percent, even large doses of ethylene are effectively antagonized. Modern practice, therefore, is to store apples for long periods in cooled, sealed atmospheres containing about 5 percent carbon dioxide. While this method is obviously beneficial in reducing apple spoilage, it has also led to the deplorable custom of almost never selling a fresh apple. The older, stored fruit is sold first, while the newly harvested fruit is placed in high carbon dioxide storage. Hence a good, fresh eating apple is hard to find; the stored product, while not rotten, lacks the distinctive flavor and aroma of the fresh fruit.

Despite its generally deleterious effects, externally applied ethylene can sometimes be useful. If hard, starchy, unripe bananas are exposed briefly to minute concentrations of ethylene, they gradually turn into the soft, sweet, speckled, ripe yellow bananas of commerce. This information has been of great benefit to banana growers and shippers. If bananas are harvested when they are soft and ripe, shipment without serious bruising injury is practically impossible; by the time the fruit arrives at its destination, it will either be completely discolored and crushed or actually rotten. Transport of the fruit became much simpler after the discovery that bananas cut off the tree while still green will ripen when exposed to ethylene. Today's growers harvest the firm, unripe fruit and ship it to storage warehouses, where, treated with ethylene, it ripens within a few days. The ripe fruit is then transported to nearby markets without injury. Most of the bananas we eat these days are handled in this manner.

If ethylene causes the artificial ripening of bananas, is it also responsible for the natural ripening of the fruit? The answer is almost certainly yes. When gas samples are removed at various times from within the fruit tissue, simple chemical analysis shows that ethylene is present and rises to a sharp peak shortly before the onset of ripening. In many fruits, this normal development of a peak of ethylene in the tissue is followed by a tremendous elevation in the rate of carbon dioxide evolution. Such a peak of respiratory gas output in fruits is referred to as a *climacteric*. Following the climacteric, changes caused by enzymes within the fruit lead to the typical softening, sweetening, and color changes. Thus, we must conclude that natural fruit ripening is controlled by ethylene produced in the plant.

Ethylene is also partly responsible for the development of the abscission layer in stalks, which causes the shedding of leaves and fruits. The unlucky chrysanthemum grower who defoliated his plants with natural gas was therefore merely hastening a natural process. Similarly, a pineapple grower who induces his plants to flower by spraying them with a synthetic hormone-type compound is simply stimulating his plants to produce extra ethylene. The same effects on flower promotion can be achieved by ethylene alone.

Saturating an entire open pineapple field with ethylene gas is obviously impractical. Normal diffusion and wind-aided convection would quickly carry the gas out of range of the plants. For this reason, growers and scientists have searched for forms of ethylene that could be applied as liquids. In the late 1960s, it was discovered that a simple, water-soluble ethylene derivative, chloroethanephosphonic acid, could release ethylene inside plant cells. A commerical version, called Ethephon, when applied in an open pineapple field at the rate of one to four pounds per acre, causes 100 percent flower induction. The sprayed fruits not only mature more rapidly than untreated controls but they also all mature simultaneously. Spraying consequently makes the pineapples more amenable to mechanical and other rationalized harvest techniques. Because of these benefits, this ethylene derivative has found wide acceptance in agriculture.

Where in the plant does ethylene originate? And why is it made at such restricted times in the plant's life cycle? Answers to these and related questions of importance to plant physiologists can now be given in rather precise terms, although that would not have been possible two or three years ago. Experiments using radioactively labeled materials show that ethylene is made from the amino acid methionine, which exists in all living cells and is a component of many important proteins present throughout the life cycle of plants. Since methionine is always present in plant cells, control over ethylene synthesis cannot depend on the availability or nonavailability of this "precursor." The control must, instead, be exerted by the enzymes that act on methionine, which must appear and disappear during the life cycle of the plant. It is now known that methionine is first "activated" by being attached to the energy-rich compound, ATP. The resulting compound, S-adenosylmethionine (referred to affectionately as SAM), can then break down by various pathways, one of which yields an unstable intermediate (ACC) that decomposes to ethylene. So, whether or not ethylene is formed from SAM depends on the presence of the appropriate enzymes for converting SAM to ACC.

One lead to identification of the factors controlling the appearance and activity of such enzymes is the discovery that the typical apical hook at the top of the stem of pea seedlings grown in the dark results from differential growth on the two sides of the hook, which is, in turn, controlled by ethylene production in the hook. Red light perceived by the plant pigment phytochrome causes the hook to open and also depresses the production of ethylene by the hook. The action of red light in facilitating hook opening can be prevented if the hook region is simultaneously supplied with ethylene from the outside. These results seem to prove that the opening of the hook is caused by the cessation of ethylene production. The inference is that the enzymes regulating ethylene synthesis are turned on and off by the phytochrome system. Whether or not auxin, known to promote ethylene synthesis, plays a role in this phenomenon is unclear.

Rotten Apples and Ripe Bananas

Hormones are usually defined as naturally occurring organic compounds, found in minute quantities in plants and animals, that produce great physiological effects when transported from their place of origin to the specific tissue on which they act. Ethylene fulfills all the requirements of a hormone, yet some scientists seem reluctant to grant it authentic hormonal status. Recognized hormones circulate in body fluids and sap through structured and specific pathways. Ethylene transport, on the other hand, seems to occur exclusively through gas phases inside or outside the plant. Whichever way this quibble ends, ethylene is firmly established as a remarkably active and sometimes useful plant growth substance.

9

Turning Plants Off and On

LAST SUMMER the vegetation outside my windows burgeoned. What with abundant rainfall and high temperatures, plant life thrived, and I had more than the usual job keeping the garden from being completely overgrown. But now the scene outside my window has changed radically. Even before the first frost approached, the plants had anticipated the winter ahead. On the apple and ash trees, leaves stopped growing and turned to brown; on other trees, scaly winter buds rather than new leaves had formed at the growing points. Meanwhile, most annuals had produced the seeds in which form they will pass the winter. All these changes, seemingly premature during the warm days of late summer, occurred because a plant's failure to effect them in good time could cause its extinction during a frost. This doesn't mean that the plant knows the date of the first frost and acts accordingly; it means that midlatitude plants have of necessity evolved appropriate timing mechanisms to insure their survival over periods unfavorable for growth. The key to the mechanism is the changing length of day.

Turning Plants Off and On

In the northern latitudes, day length reaches its maximum on about June 21 and its minimum six months later. September 23 and March 21 are the equinoxes: times when day and night are of equal length everywhere on earth. At the latitude of my home near New Haven, Connecticut, day length has diminished from its maximum of about fifteen hours to twelve hours by the end of September. It will be reduced to nine hours at Christmas time. Plant leaves take account of this change through the blue pigment phytochrome, which manifests one form during day and another form at night. As the length of day shortens, this pigment exists less and less in its daytime form and more and more in its night form. In the meantime, the plant measures the passage of each 24-hour period through its endogenous circadian, or internal daily, rhythms. But how can the interplay of phytochrome and rhythms in the leaf affect the behavior of the bud at the growing point, perhaps many feet away?

About a decade ago, Philip Wareing and some of his students at the University College of Wales at Aberystwyth studied the effect of diminishing the daily photoperiod, or light exposure, on the formation of winter buds in the birch tree. They noted that, as the day length shortened, the rate of stem elongation at the apex decreased and eventually ceased altogether. At the same time, the tree stopped making full-sized new leaves; instead, small, scaly leaves began to surround the now dormant bud. They reasoned that some sort of growth inhibitor accumulating at the stem apex might be causing these changes. By making appropriate extracts of the bud as the growing season progressed and testing them on other living plant tissues, the researchers did indeed find an increasing concentration of a substance inhibitory to growth. When purified and applied to plant tissues, this substance not only caused a

WOODY SHOOTS IN WINTER

Magnolia

Butternut

Norway
Maple

Staghorn
Sumac

In the winter, twigs of staghorn sumac, magnolia, Norway maple, and butternut lack expanded leaves. Tiny leaves for the next year's growth are encased in winter buds, with internal insulation, tightly overlapping bud scales, and external "varnish."

marked diminution in growth rate but also partially mimicked the effect of short days in promoting winter bud formation. When applied to viable seeds, it created a dormancy, which could be overcome by washing away the applied material. In other words, this inhibitory chemical, produced under the influence of artificially simulated short days, seemed to induce in the plant all the well-known effects of the natural short day itself.

Through cooperation with chemists, Wareing and his colleagues succeeded in isolating pure samples of the effective material. It was then only a matter of time until the chemical nature of its molecule was determined, and this was followed shortly by the development of a purely synthetic method for the production of any desired quantity of the inhibitor. Because the material not only induced the formation of dormant buds but caused seeds to become dormant as well, it was given the common name *dormin*. (Its proper chemical name is much more complicated.)

Scientists are aware that an idea, "ripe" for discovery, is frequently uncovered virtually simultaneously by several different persons working independently of each other. So it was with dormin. At Davis, California, thousands of miles from Aberystwyth, Frederick Addicott and several colleagues were working on the problem of the premature shedding of cotton bolls. This anomaly in certain high-yielding varieties of cotton plants caused them to lose their bolls spontaneously shortly before harvest. Addicott felt that the aberration might be triggered by the movement of a substance from the leaf to the boll stalk. Like Wareing, he made extracts of slightly senescent leaves, applied the substance to test systems, and succeeded in purifying, isolating, and identifying the effective material. Because it caused shedding, or abscission, he called it *abscisin*.

Wareing and Addicott published their results separately,

and to their common surprise, they learned they had isolated the same chemical material. Because Addicott's report was published one week before Wareing's, the compound, which is actually an organic acid, was ultimately named *abscisic acid*. ABA, as it is known to botanists, has become an important substance for the plant physiologist and the agriculturist, as it seems to have remarkably versatile effects on many aspects of plant development. Not only does it slow down the over-all rate of growth and predispose the growing points toward a dormant state, it probably plays an important role in the regulatory physiology of plants.

Recall that water loss from leaves, in the form of vapor diffusing through open stomata, is one of the major threats to a plant's continued existence. If this evaporation is too extensive, the leaf can wilt, desiccate, and die. Extreme damage is sometimes prevented because wilting leads to loss of turgor in the guard cells surrounding the stomata. When these cells become limp, the stomatal pore closes, and further water loss is greatly reduced or entirely stopped. It is in this process, the closure of the stomatal pore, that abscisic acid plays its important role.

Experimenters at Wye College in England recently analyzed both turgid and wilted leaves for their abscisic acid content. They found that wilted leaves with closed stomata—a protection against further water loss—had consistently high levels of ABA, often ten to twenty times the normal concentration. When ABA was applied to a normal leaf with open stomata, the stomata closed as if the leaf were wilted. This closure was brought about through loss of potassium and, thus, water from the guard cells. The wilting-induced increase in concentration of ABA therefore prevents additional water loss from a partly desiccated leaf and could save the plant's life.

This dehydrating effect of abscisic acid may also help

explain another extremely puzzling aspect of a plant's preparation for dormancy. Normal plant cells contain about 90 percent water, while seeds tend to have about 10 to 20 percent. How can a seed be dried out while still attached to the wetter mother plant? It appears that ABA may again be involved. The movement of large quantities of ABA into the seed causes it to lose water to surrounding tissues, even though the over-all water content there may be greater. The desiccated seed remains dormant over the winter and does not start to germinate until the level of ABA declines. In fact, the amount of ABA initially incorporated into the seed and the rate at which this quantity declines over the winter may constitute the chemical time-keeping device that tells the seed that the last frost day has passed and it is safe to germinate. The dormant winter buds of trees and shrubs may use the same device. Thus, ABA may be considered the "turn off" switch that converts an active plant into a dormant one. This is a handy, indeed necessary, survival device in the midlatitude zone.

The ability of ABA to close stomata and thereby protect plants against excessive water loss can perhaps be turned to practical agricultural advantage in arid zones. The problem will be to effect closure of the stomata without simultaneously overheating the plant or converting actively growing buds to the dormant state. A solution to this problem could have great importance for a world faced with food shortages, especially in those countries that must try to grow crops more efficiently in suboptimal environments.

10

Sex and the Soybean

THE SOYBEAN is not innocent of sex. Unlike higher animals, which carry formed sex organs throughout their lives and use them only sporadically for reproduction, plants make their sex organs, flowers, only shortly before reproducing. Plants form flowers at a time during the annual weather cycle that will guarantee successful production of progeny. Like other higher plants, the soybean has adopted a variety of biological mechanisms to insure that its flowers form at the right time of year.

For most of its life, the soybean vegetates, producing only roots, stems, and leaves. Through the process of photosynthesis, the plant nourishes itself, making carbohydrate from carbon dioxide in the air and water in the soil, with the aid of sunlight absorbed by chlorophyll and other pigments in bodies called chloroplasts. Each cell of the soybean leaf has between fifty and one hundred of these active green bodies, and the total amount of carbon fixed can be prodigious. By cooperating with *Rhizobium*, a bacterium present in almost all fertile soils, the roots of the soybean plant are able to fix atmospheric nitrogen gas, converting it to ammonia, which then finds its way into amino acids and proteins. This process of nitrogen fixation occurs in nodules, warty protuberances of the root formed after *Rhizobium* cells invade root hairs. Neither the soybean plant nor the bacterium is capable of fixing nitrogen on its own; only the symbiotic association of the two in the nodule makes this process possible.

Each active *Rhizobium* contains a small circle of deoxyribonucleic acid (DNA) in addition to its regular circular chromosome; this smaller circle, called a plasmid, contains the genes for nitrogen fixation (*nif*), as well as other genes controlling nodule growth. Introduced into the soybean root cell when the bacteria invade through a root hair, the *nif* genes become active and synthesize the enzymes that make nitrogen fixation possible. This tidy cooperation enables the soybean plant to grow independent of nitrogenous fertilizer and even to enrich the nitrogen supply of the soil. So sensitive is the symbiotic nitrogen fixation system that it turns off when nitrogenous fertilizer is added to the soil, as if the plant switches to conserve energy when already fixed nitrogen is available.

With its food assured through photosynthesis and its nitrogen supplied through direct fixation, the soybean plant is well on its way nutritionally. All it needs is water and mineral elements such as potassium and phosphate; these are usually available in a fertile soil or can be easily and inexpensively supplied in fertilizer. These properties, together with a high bean yield, have made the soybean harvest one of America's most bountiful and the soybean one of the most profitable export crops. Virtually nonexistent in the United States in 1940, soybean acreage now about equals that of corn.

But the soybean plant cannot complete its life cycle without a dramatic change from the vegetative habit. The plant is of little use as a crop until it produces beans, and to make beans it must first make flowers. Like all life processes, a plant's transition from the vegetative to the sexual habit—flowering—is genetically determined. Some varieties of soybeans, such as *Agate* or *Batorawka*, will flower when they reach a certain age or stage of maturity without relying on environmental conditions. Others, such as *Peking* or *Biloxi*, require short days before they can initiate their sexual stage. Each short-day type of soybean has a certain critical photoperiod, a length of day beyond which it will not flower.

Thus, as the days become progressively shorter in the fall, a date just below the critical day length eventually arrives and the soybean flowers. This photoperiodic requirement insures that the plant will be large and vigorous before it flowers; presumably, it has grown and vegetated from May to August, and now has enough leaves, root nodules, and other vegetative organs to support a large crop of fruit. The date of the photoperiod also guarantees that flowering and fruit-set will be complete before the first frost. There are soybean varieties with photoperiods to suit a wide diversity of agricultural regions.

The photoperiodic control of flowering was discovered in the United States in 1920, but despite more than half a century of intensive research all over the world, the intimate biochemical details of the process are still not understood. Since the late 1930s, scientists have known that the soybean uses the dark rather than the light period of the day to measure its photoperiod. Thus, if the critical photoperiod turns out to be fourteen hours, the true control is ten hours of unbroken darkness. The critical dark period is effective whether it is surrounded by four or fourteen hours of light. To some degree, dark-time measurement is controlled by the state of phytochrome, a pigment discovered in 1959 that governs the plant's response to light.

Sex and the Soybean

When exposed to sunlight, this pigment becomes active; when stored in darkness, it decays back down to its original, or ground-level, state. Another component of photoperiodic timing, besides the hourglass scheme of phytochrome activation and decay, is the still-unknown generator of biological rhythms.

Although the buds of the plant change suddenly from producing leaves to forming flowers, the vegetative leaves of the plant, rather than the buds, perceive the photoperiodic stimulus. A short-day Biloxi soybean that has been kept vegetative by exposure to long days can be made to flower by the simple trick of tying a black bag over the leaves for enough hours to add up to an uninterrupted long night of at least the critical length. Even if all the plant's leaves except one are trimmed off, exposure of the remaining single leaf to the right photoperiod will result in prompt conversion of the distant bud to the flowering habit. This experiment indicates that some substance formed in the leaf moves to the bud and influences its behavior. This substance, the hypothetical reproductive, or sex, hormone of plants, has been given the name *florigen*, a hybrid term formed by combining the Latin word for "flower" with the Greek term for "giving rise to."

Simple grafting experiments confirm the existence of florigen. If a flowering plant of Biloxi soybean is grafted onto a vegetative plant that has been maintained on long day, then both the receptor and the donor will develop numerous flower buds. For this experiment to succeed, there must be a true union of tissue between donor and receptor, implying that florigen can move only through living cells. By varying the duration of living-cell contact between donor and receptor, the experimenter can calculate that the floral stimulus moves from leaf to bud through the stem at a rate of one to several centimeters per hour,

(a) Soybeans on long days produce no flowers.

(b) When exposed to four successive short days, the plant initiates flowers, visible after several days.

(c) When the flowering plant is grafted onto a vegetative plant maintained on long days, the latter is also induced to flower. This indicates that a mobile substance passes between the donor and the receptor plant.

or approximately the rate at which sugar moves through the phloem tissue of the plant's conducting system. If the experiment is done well, then a flowering short-day plant can induce a vegetative long-day plant to flower, and vice versa. Florigen appears to be common to all types of flowering plants.

The graft transmission of florigen is most dramatically displayed when only a single induced leaf is grafted from a donor to a receptor plant, the union forming along the diagonally bisected leafstalk, or petiole. This single donor leaf can be effective even if only a small fraction of its leaf blade area remains. The floral hormone must be a very effective compound, since only very tiny quantities of material could move from the small donor leaf tab through the plant to the distant bud.

The sensitivity of the leaf system regulating florigen synthesis is shown by the fact that only a momentary flash of light during the long dark period is enough to completely interrupt the flowering process. This tiny amount of light energy, perceived by phytochrome, is not enough to grossly transform many molecules through direct photochemical effects. Hence, a very little of some effective substance must be changed. Even the intensity of bright moonlight may alter the flowering response, which gives some support to old theories of planting by the light of the full moon. At the next lunar cycle, a long-day plant would be made to flower as a result of the extra bright moonlight, while the flowering of a short-day plant would be prevented or partially inhibited.

One curious bit of quantitative hocus-pocus is especially baffling. Although a few minutes difference in day length —actually night length—can make a complete difference in flowering behavior, one single photoinductive day length is not enough to cause flowering in soybeans, even though

Sex and the Soybean

it is enough in some other plants. For floral initiation, soybeans require a minimum of four successive short days. If two short days are followed by one long day and then two more short days, the arithmetic is incorrect for the soybean; four correct short days must follow without interruption. Perhaps each successive correct photoperiod leads to the synthesis of a little bit of florigen that is still not enough to cause the plant to change over to the sexual habit; only four correct photoperiods allow the buildup of the required amount of florigen. Perhaps florigen is somewhat unstable, decaying rapidly under unfavorable photoperiodic conditions. If so, then even if three successive photoperiods generated almost enough florigen to initiate flowering, this quantity would all disappear after only one unfavorable photoperiod, and the plant would have to start over to receive its four correct day lengths in a row. Once the critical level of florigen is attained, however, the florigen-producing reaction is self-sustaining.

The physiological evidence for the existence of florigen is so sound that presumably it could be isolated from masses of properly induced leaves, those exposed to the correct photoperiod. But despite many attempts by able investigators over the past forty years, no unambiguous success has been achieved. In the interim, numerous other plant hormones regulating growth—including auxins, gibberellins, cytokinins, abscisic acid, and ethylene—have been isolated and characterized. Why the special difficulty with florigen? There are several possible answers: florigen may be difficult to extract, unstable once extracted, or difficult to reintroduce properly. Again, florigen may not be a single substance, like the other hormones, but a mixture of substances that are separated from each other in the leaf extract during the purification procedure. Florigen transport does seem to follow sugar-transport patterns in the plant, and the negative effects of noninduced leaves (not given a photoperiod that leads to flowering) are manifest only if they are located between the florigen-producing leaves and the bud to be influenced. Therefore sucrose, the major transport sugar of plants, could have an influence on flowering habit. In some plants that have remained vegetative because of incorrect photoperiod, simple defoliation can cause flowers to appear. The leaves may have inhibited flowering because their export of sucrose to the bud blocked the transport of florigen to an induced leaf.

The hunt for florigenically active chemicals proceeds in many laboratories, partly because of the potential practical importance of such substances. As we know, pineapples can be chemically induced to flower by ethylene or ethylene-generating systems. Similarly, gibberellins can induce some long-day plants to flower. In tissue culture experiments, sucrose is sometimes an important factor in the appearance of flower buds. But the true nature of elusive, transmissible, universally active florigen remains mysterious.

11

Plants Have a
Few Tricks, Too

ALL CREATURES living in the wild are subject to attack by predators, and their survival as a species depends in large measure on their success in fending off such attacks. Animals have many obvious self-defense mechanisms. Some, for example, hide from enemies by merging into the landscape so that it is difficult to see them; others hide by deliberately popping into a hole or under a rock. Because of their speed, some creatures outrun potential predators, while others outlast them by superior stamina during a long pursuit. A cornered or alarmed animal can ward off an enemy by obnoxious odors, gestures, or noises, and in a pinch can stand and fight off an attacker. Even when faced with a microscopic invader, such as a bacterium, a fungus, or a foreign protein of any kind, many animals can react with a variety of defenses, including ingestion of the attacker by special mobile body cells or the formation of specific antibodies that couple with the invading protein or cell and render it harmless.

Plants, by contrast, seem at first sight to be relatively defenseless against attack, but a closer look reveals that they do have some mechanisms for warding off attack by other organisms. For the most part, they are incapable of sufficiently rapid motion to do damage to an animal, although the insect-trapping devices of the sundew (*Drosera*) and of Venus's-flytrap (*Dionaea*) have been widely popularized. Upon mechanical stimulation the leaves of the sensitive plant (*Mimosa*) will rapidly fold, which is said to protect the plant against foraging animals. Thus, botanical humor has it that a goat entering a patch of wild *Mimosa pudica* would starve to death, because the plant's sudden folding of its leaves after jostling would make it seem unavailable for foraging. I personally doubt whether such an obvious ploy would deter an omnivorous goat.

Another kind of movement results from the tactile sensitivity and coiling growth of tendrils and other climbing organs; this permits some vines to grow over trees and, as in the case of the strangler fig, to completely kill the more upright host. The tree can do nothing to escape the ever tightening clutches of its unwanted epiphyte.

Most successful plant defenses are exerted against insects and microbes. It is well known, for example, that among closely related varieties or species of plants, some are eaten by insects while others are not, and that some are susceptible to a disease while others are not. In such cases the differences between the related plants are often a clue to their defense mechanisms. Protection may be rather mechanical; some leaves are very leathery in texture and are covered on both surfaces by a waxy cuticle or a thick, cushiony tuft of matted hairs. Such structural modifications of the leaf's surface repel some insect predators much as thorns repel some animals.

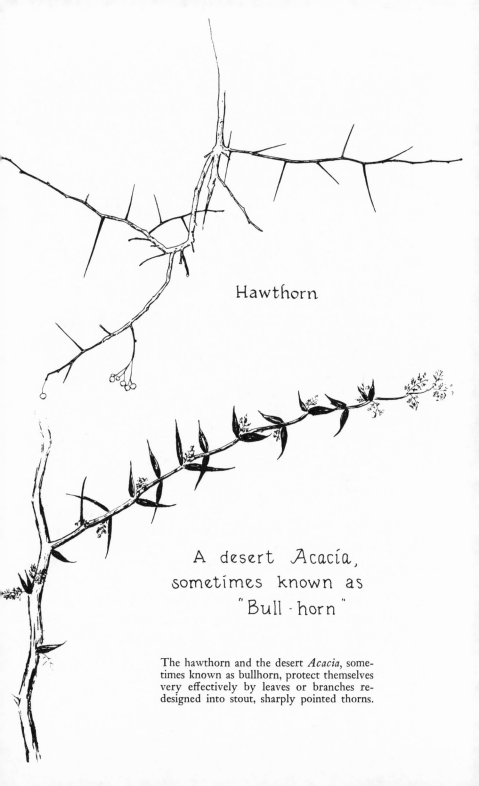

Hawthorn

A desert *Acacia*,
sometimes known as
"Bull-horn"

The hawthorn and the desert *Acacia*, sometimes known as bullhorn, protect themselves
very effectively by leaves or branches re-
designed into stout, sharply pointed thorns.

Other defense mechanisms are chemical. For example, many wild plants contain bitter-tasting chemicals like alkaloids, tannins, or simple phenols, whose value to the plant is not well defined. Because of a general belief that everything in a wild creature must have some function (or else it would have been selected against and eliminated during the course of evolution), it has been suggested that these materials may discourage insects and large animals from eating the plant. Similarly, pungent, volatile materials like those of the onion and mustard are said to repel some insects at a distance, before they even get to the plant.

In some instances plant pathologists have been able to draw correlations between a plant's content of certain chemical components, such as the phenols, and its resistance to fungal diseases. Phenols are, after all, well-known germicidal materials; the carbolic acid so easily smelled in hospital corridors is phenol itself. If effective against external microbes, why not against internal ones as well?

But there are some difficulties with this theory. Why don't the phenols kill the plant itself? The answer may be that the phenols, tannins, and other germicidal materials of plants are found in vacuoles, separated from the living part of the cell by a membrane through which they cannot pass. They do not, therefore, act to repel or kill an invader unless the cell is first attacked in such a way as to break the membrane down and "liberate" the previously restricted phenol. The invader thus triggers the release of a counterweapon hidden in a storage vault in the cell.

In recent years, a more active defense against microbial invaders has been shown to exist in some plants. When attacked by filamentous molds, these plants respond by making a germicidal compound that they did not contain before the invasion. These substances are called phytoalexins, from the Greek *phyton*, "plant," and *alexin*, a warding-off substance. Unlike antibodies, they tend not to be specific with regard to fungal toxicity, and are restricted to a zone immediately surrounding the infected area. Thus, they are of no use in providing systemic immunity.

One phytoalexin, a complex phenol called pisatin, has been isolated from pea pods inoculated with the pathogenic fungus *Ascochyta* or the nonpathogenic fungus *Monilinia*. In general, pea varieties resistant to the pathogenic organism form more pisatin than nonresistant varieties, a picture consistent with a functional role for pisatin in disease resistance. Some especially virulent invaders have the ability to break down the pisatin that is formed by the plant, which may account for their virulence. Plant and invader appear to deliver thrust and counterthrust in the chemical battle for survival.

Tissues invaded by filamentous fungi also seem to form large quantities of certain oxidative enzymes, like peroxidase. When peroxidase acts on phenols, it converts them to "free radicals," especially active forms of these compounds that may be the actual toxic material acting against the fungi. Thus, the active defense mechanism of the plant may involve not only the formation of a potential chemical toxin but also of the catalytic "fuse" that activates it.

There is much interest in these recent findings among plant geneticists, for the production of new agriculturally important and disease-resistant crops may be linked to phytoalexin production and activation. If successful, such an approach might even lessen our dependence on troublesome, externally applied pesticides.

The Plant Moves
About

12

The Language of
the Leaves

FOR THOSE who would divine some of the inner secrets of plants, the leaves speak a language of their own. Learning to read and speak this language is a challenge worthy of the most highly developed skills of the experimental biologist; learning to frame critical questions and ask them of the plant in such a way as to elicit a clear answer is exhilarating. I speak from experience, for some colleagues and I have been engaged in this kind of dialogue for about ten years now, and the answers we receive from the plant keep us busy devising ways to ask still more questions.

Let us start with the observation that many leaves have obvious sleep movements: during the day, the leaves are so oriented that the lamina, or blade, is roughly perpendicular to the incident sunlight, and is thus said to be in an upright, or "awake," condition. By contrast, at night the leaves frequently fold so that their blades face either other leaf blades or the stem of the plant. It is easy to understand why leaves should be oriented with their blades perpendicular to the incident light during the daytime: the leaves are, above all, photosynthetic organs and the horizontal orientation optimizes light absorption for this important nutritional process. But why should a leaf fold at night? Why go through the trouble of a cycle of closing and opening when it would be less complicated simply to stay open and be ready to receive the next morning's sunlight?

Charles Darwin, who was fascinated by plant movements in general, and especially by leaf movements, hypothesized that the inward folding of leaves at night prevented excessive radiative heat loss to the open sky. Since, on a clear night, the open sky is a "black-body" absorber with an effective temperature of absolute zero, and since heat loss by radiation is proportional to the fourth power of the absolute temperature difference between sending and receiving body, such heat loss could cool the plant down drastically. For tropical plants, whose metabolic processes are geared to work best at high temperatures, this could be a serious impediment to growth. Thus, folding of the leaves at night, according to Darwin, is simply a device employed by the plant to protect itself against excessive radiative heat loss.

For about 80 years this was the only theory that seemed at all reasonable, and it was accepted simply for lack of an alternate explanation. Recently, however, another plausible theory has been advanced by the distinguished German plant physiologist Erwin Bünning. Bünning noted that a necessary stimulus to flower formation in many plants with pronounced leaf movements is the so-called short-day treatment. The plant is so sensitive to light during the mandatory dark period of the short day that even the intensity of bright moonlight could inhibit production of the floral hormone, florigen.

How, then, can such photoperiodically sensitive plants

time their flowering efficiently when the erratic effects of moonlight might disturb them? The answer, said Bünning, is that the leaf-folding process causes the plant to present the narrow edge of its blades to the impinging moonlight; this minimizes light absorption to the point that the chemical processes requiring darkness can proceed without interruption. In long-day plants, which require short nights for flower formation, the same mechanism would prevent premature reproduction.

Which theory is more nearly correct, Darwin's or Bünning's? It is hard to say, although Bünning's has an appeal in that it bases the plant's movements on a more qualitatively important decision. To make a flower or not seems to be a more important question than whether to cool down markedly or not. For some tropical plants the latter decision may be crucial, but even many temperate plants have pronounced nyctinastic (sleep) movements. At the moment, it seems most appropriate to concede that both theories have a certain validity.

What is the mechanism that causes the leaves to move? Certainly not muscles, which do not exist in plants, although contractile proteins do exist within plant cells. This question is most conveniently studied in certain leguminous plants whose leaves are attached to the stem by a special organ called a pulvinus. This swollen knob of tissue contains special motor cells on its upper and lower surfaces; such cells are capable of shrinking and swelling dramatically in response to loss and gain of water. This, in turn, is governed through regulation of the salt concentration of the storage compartments, or vacuoles, of these cells. When the salt concentration of these cells is increased, their osmotic concentration becomes sufficiently high that water tends to diffuse into the cell, thereby increasing its internal turgor pressure and making it bulge and swell. When salt is

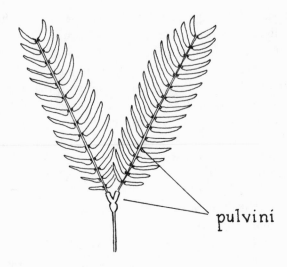

pulvini

Leaflets of *Albizzia*, showing the *pulvini* by which the individual leaflets and the collection of leaflets are attached to the main petiole.

lost from such motor cells, the osmotic concentration declines and water tends to diffuse out, the internal pressure declines and the cell becomes flaccid and shrinks.

Thus, when the upper pulvinal cells of certain plants are turgid and the lower cells flaccid, the leaves are pushed down and out from the stem to the horizontal, or "awake," position. When, on the contrary, the lower cells are more turgid than the upper cells, the leaf is pivoted at the pulvinus and pushed upward and toward the stem, assuming the closed, or "asleep," attitude. Thus, leaf movements are accomplished by differential salt accumulation patterns, which in turn control the diffusion of water into and out of key motor cells in the pulvinus.

What controls salt movement into and out of the motor cells? One obvious answer is the alternation of light and dark. In fact, one can easily show that many darkened, asleep leaves will open promptly upon illumination; the opening is accompanied by a marked increase in the salt concentration of the upper motor cells. Conversely, many illuminated, awake plant leaves will close promptly upon transfer to darkness, and this is accompanied by a loss of salts from the upper motor cells and a gain of salts in the lower motor cells.

In the plant we have studied most intensively, the leguminous silk tree, *Albizzia*, the mobile element is potassium, probably moving across the membrane as a positively charged ion (K^+), accompanied or followed by the negative chloride ion (Cl^-). The uptake of potassium salts from upper motor cells of the pulvinus involves the expenditure of energy, indicating the operation of a metabolically powered "pump" located in the membrane. The pump uses the energy of ATP to expel a proton or hydrogen ion (H^+); K^+ then moves into the cell to replace the lost positive charge. The reverse process occurs when K^+ is lost from the opposite pulvinal cell.

But this is far from the whole story. If a leaf is artificially maintained in continuous light for an unusually long period, the leaf closes, even in bright light. Conversely, if a closed leaf is maintained in an extraordinarily long dark period, it will open spontaneously. The closing and opening movements of leaves maintained under constant conditions will occur at roughly twelve-hour intervals, so that the complete cycle takes about 24 hours, or one day. Such rhythmically repeated events, with an oscillatory frequency of about one day, are said to be circadian. When the circadian rhythm persists under constant environmental conditions, it is said to be endogenous. Thus, the leaf movements of many plants show control both by light and by some as yet uncharacterized endogenous, circadian process.

Albizzia leaves - open & closed

Albizzia leaves, open and closed. The transition may be caused by transfer of leaf from light to dark, or by the operation of an endogenous rhythm.

The Language of the Leaves

Is the rhythmic movement the same as the light-controlled movement? We have found that endogenous closure of leaves maintained in the light differs from nyctinastic closure caused by transfer of leaves from light to dark. While the latter process involves, as mentioned above, the operation of an energy-requiring, active pump, endogenous closure in light can occur even in the presence of substances that prevent such active processes from occurring. Rhythmic closure therefore seems to involve a passive leakage outward of the potassium salts that were previously energetically accumulated in the upper motor cells. This fact led us to hypothesize that the rhythm involves an alternation between two states of the differentially permeable membranes of the motor cells, a state of "integrity" and a state of "leakiness." During the leaky phase, salts leave the cell no matter how rapidly the inwardly directed active pump is working; when the membranes regain their integrity, salt accumulation again predominates.

We have no definite proof that this hypothesis is correct, nor do we have any reasonable mechanism to propose for the rhythmically changed states of the membrane. But because the interplay of light and rhythms is important in the control of a myriad of plant processes, including flowering, we are pressing our investigations further. How can light affect membrane permeability? Can chemical substances control these movements? How can changes in leaf cell permeability affect behavior of plant parts at a distance? The answer to these questions could explain not only the sleep movements of leaves, but also flowering, senescence, hormone action, and a host of other problems.

13

A Basic Unity

of Life

IN THE LAST CHAPTER, I discussed the probable significance of, and the basic physiological processes controlling, the sleep movements of leaves. Now I want to focus on the biochemical mechanisms involved in the leaf movements of the leguminous silk tree, *Albizzia*. The leaves of this plant are subdivided into pairs of opposing leaflets, which lie open during the day and close tightly against each other at night.

The main chemical event regulating the movement of these leaflets is the movement of potassium into and out of the motor cells of the pulvinus, the fleshy joint connecting leaf stalk and stem. The leaflets are open when the upper cells of the pulvinus are rich in potassium. The potassium richness raises the osmotic concentration of the upper cells; they thus take up water and swell, forcing the leaflet to pivot down and away from the stem. When the upper pulvinal cells lose potassium and the lower cells gain it, the opposite changes occur, and the leaflets fold together, or close. The leaflets are generally open in the light and closed in the dark, although the major control is an internal clock that seems to work by controlling the leakiness of the pulvinal cell membranes to potassium and other solutes.

Whether or not a leaflet closes on transfer from light to darkness depends in the first instance on the phase of the internal rhythm. Erwin Bünning suggested that the "photophile," or light-loving, half of the cycle begins when the light goes on, and that the "scotophile," or dark-loving, phase starts twelve hours later. Thus, a leaf that has been in light for, let us say, ten or eleven hours will close promptly upon being darkened, because each half of the circadian (daylong) rhythm lasts for about twelve hours. Even if left in light for more than twelve hours, such a leaf will start to close, but if transferred to darkness it will close much more rapidly and completely.

On the other hand, a leaf transferred to darkness after only one or two hours of light will not close very vigorously or completely. This is "explained" by saying that the clock that keeps time for the plant in the measurement of the daily rhythm is still in the photophile setting, and is thus not in the correct state for dark reactions such as closing. This is not really an explanation, but simply another way of saying that the rhythm can be shown to exist, and can be measured and even defined in terms of the changing ability of leaflets to fold together in darkness after having received varying periods of light. The 24-hour rhythm of terrestrial plants has presumably evolved in response to the earth's 24-hour day. Thus, if plants are ever discovered on other planets with different day lengths, they could be expected to have differently timed rhythms if they have rhythms at all.

All plants possess small quantities of phytochrome, a remarkable pigment that is very sensitive to light. When a plant is stored in darkness for several hours, the bulk, if not all, of its phytochrome is in a form that absorbs red light strongly. Upon irradiation with ordinary sunlight or any sort of mixed wavelength "white" light, the pigment is transformed to another molecular configuration, which no longer absorbs heavily in the red region of the spectrum, but rather in what we call "far-red," or just barely visible longer wavelength light. During daylight, which contains more energy in the red than in the far-red part of the spectrum, phytochrome is kept predominantly in the far-red absorbing form. In darkness, this form is spontaneously transformed to the red-absorbing form. For convenience, the red- and far-red-absorbing forms of phytochrome are called P_r and P_{fr}. The transformation of P_r to P_{fr} by light —and the subsequent reversion of P_{fr} to P_r in the dark— seemed to provide an obvious chemical "hourglass" mechanism that could furnish a basis for the alternating photophile and scotophile phases of a plant's behavior. Further investigations have shown that the timing mechanism is, in fact, much more complex.

The change in absorption properties of phytochrome as it changes from P_r to P_{fr} provides the laboratory experimenter with a convenient handle. Because light, to be effective in producing any chemical change, must first be trapped by an appropriate absorbing pigment, phytochrome provides a convenient two-way switch for turning off and on any process it controls. If conversion of phytochrome from the P_r to the P_{fr} form is responsible for the opening of the leaflets as they are moved from dark to light, then red light should cause opening and far-red light given immediately after the red light should prevent it from acting. Although the state of phytochrome is not the primary regulator of leaflet *opening*, it does closely regulate the *closure* of leaflets when they are transferred from light to dark.

A Basic Unity of Life

This relationship can be shown by the following simple experiment. If after the initial, natural light period (phytochrome is now present as P_{fr}), the plant is briefly irradiated with far-red light (P_{fr} will now be reconverted to P_r) and then put into darkness, the leaflets will not close rapidly. If the far-red light is omitted before transfer to darkness, or if red light is administered after the far-red and then the plant is put into the dark, the leaves will close promptly and vigorously. Thus, phytochrome can control leaflet closure, but only if the circadian rhythm permits any closure at all. The rhythm may be thought of as controlling a master valve and phytochrome as controlling a subsidiary valve that works only if the master valve is open.

Recalling that leaflet movement depends ultimately on the transport of potassium from one set of pulvinal cells to another, we can now ask how phytochrome regulates potassium movement. One provocative suggestion, put forth by Mark Jaffe, is that phytochrome somehow regulates the quantity or release of acetylcholine at the membrane of the motor cells of the pulvinus, and that this, in turn, accounts for the movement of potassium out of a cell. This mechanism is clearly borrowed from the animal world, for it is known that acetylcholine is one of the substances that transmit stimuli from one nerve cell to another across the synapse, or gap separating the two cells. Because transmission of neural impulses is correlated with changes in the sodium and potassium balance of the cells, the proposed mechanism would appear to be logical. The postulate was also supported by the fact that root tips can be shown to change their surface electrical charges reversibly upon irradiation with red and far-red light, leading them to adhere to, or be released from, negatively charged surfaces. This change in surface charge can be related to secretion of positive ions like potassium.

Using the beating of an excised clam heart to measure

the amount of acetylcholine in plant extracts, Jaffe was able to find correlations between the amount of the substance and the state of phytochrome. In support of his views, several other investigators found that the application of acetylcholine to plant tissues can cause effects that mimic the action of light absorbed by phytochrome. But the issue is not yet settled, for still other workers, including some in my own laboratory, have come up with some negative evidence. The answer to this problem will probably not be definitely obtained until new kinds of experiments are devised.

Acetylcholine is not the only neurotransmitter substance that could be involved in this plant response. Not only does acetylcholine exist in plants, together with the enzymes that make it and break it down, but so do other active neurotransmitter substances, such as serotonin, epinephrine, and norepinephrine. We are currently investigating their possible role in leaf movements and other rapid plant responses. Attention has also recently focused on a substance called cyclic adenylic acid, which acts as a "secondary hormone" messenger in many animal responses and appears to be able to elicit numerous responses in plant tissues as well. Finally, the fact that the well-known plant growth hormone indoleacetic acid is chemically closely related to serotonin has not escaped notice.

It would be poetic justice if substances discovered by animal biochemists and pharmacologists were to play an important role in solving problems in the physiology of plants, whose numerous products—such as atropine, curare, LSD, and digitalis—have been of such aid to animal physiology and medicine. And if one of the neurotransmitter substances of animals should prove to explain plant leaf movements, the unity of basic biological mechanisms would be further emphasized.

14

The Limits of
Plant Power

I DO NOT BELIEVE that plants respond electrically to human thoughts and emotions or to distant traumatic events happening to other organisms. But that doesn't mean that I doubt the existence of electrical manifestations in plants. Indeed, this old and respected aspect of plant physiology has always fascinated me. Colleagues working in my laboratory at Yale University have recently done some elegant experiments that link certain plant movements to measurable electrical changes in the tissues. The time seems ripe for new advances in our understanding of plant electrophysiology.

The electrical activities of plants are not nearly as large or spectacular as analogous phenomena in animals. The electric eel, for example, can produce several hundred volts with an output of about 100 watts. This is enough to kill a small animal, stun a man, or light a series of bulbs. No plant can do this or anything close to it. The usual plant potentials of about 100 millivolts are more than a thousandfold less than those of the electric eel, and plant currents of several microamperes are also quite low when compared with those of animals. Animal response to externally applied current is also much greater than that of plants. Thus, a brief electrical pulse administered to a nerve can cause violent twitching in an attached muscle (witness the famous frog leg experiment of the Italian physiologist Luigi Galvani, performed in the late eighteenth century), while a similar stimulus given to a plant results in a gentle and slow change in the growth rate of the stem or root or possibly in a slow growth curvature in the direction of the positive or negative electrode.

The faster and larger animal responses are undoubtedly due to the presence of nerves, those specialized cells whose entire structure and physiology are geared to the production and conduction of electrical signals. The transfer of any stimulus along a nerve causes electrical changes along the surface of that cell, and "action potentials" (voltage changes in response to an applied stimulus) of ten or more millivolts are readily recorded. Both the propagated impulse and the recorded potential arise because of changes that occur across the membrane of the nerve cell; they, in turn, lead to a flow of local currents along the outside of the nerve.

The membrane changes that cause the observed potentials and current flow are connected with movement across the membrane of certain salts, especially of charged ions like sodium and potassium. Outside the human nerve cell, the major ions are positively charged sodium and negatively charged chloride, the two components of ordinary table salt. Inside the nerve cell, sodium is a minor constituent and is largely replaced by potassium, a related element. The unique permeability characteristics of the nerve membrane permit the separation of these highly mobile salts into two compartments, the inside and the outside of the nerve cell. Potassium is roughly fifty times more concentrated inside the cell, while sodium is more concentrated outside. This unequal distribution of dissolved and freely diffusible ions across the membrane is what produces the observed electrical potential. If, for example, there is a tenfold difference in the external and internal concentrations of a salt, the resultant potential is about sixty millivolts, a value frequently observed in biological systems.

When an impulse is propagated along a nerve, a wave of "depolarization" occurs, meaning the observed potentials will suddenly drop. The drop is due to the opening of pores in the membrane, permitting the free diffusion of sodium into the cell and potassium out of it. This exchange diminishes the gradients for both sodium and potassium, and the diminished potential across the membrane is the result. During the recovery phase, the operation of "ion pumps" in the membrane restores the original differences in concentration. These pumps, made of protein, utilize metabolic energy to move ions against their natural gradient. This latter process, known as active transport, is common to all living cells.

In plants, fairly large electrical signals are manifested during vigorous movements, such as the folding of the leaflets of a *Mimosa* ("the sensitive plant") when they are touched or the sudden closing of the Venus's-flytrap. The folding together of certain leaves during sleep, which results from pressure changes in the motor cells at the base of the leaf stalk, are known to be due to cellular changes in the content of salt, especially of potassium. Recently, plant physiologists Richard Racusen and Ruth Satter implanted electrodes into these motor cells and detected electrical potential differences between their dorsal and ventral halves. The observed signals changed in an orderly way during a daily cycle, and the oscillations in potential showed some correlation with both the internal daily rhythm and the light-dark transitions in the outside world. An analysis of the mechanism of the movements exhibited some parallels with nerves, in that depolarization, ion movement through pores, and ion pumps were all involved.

Many years earlier, biophysicists E. J. Lund at the University of Texas and Harold S. Burr at Yale had shown that electrical fields and gradients could be mapped in almost all organisms, including many plants. Lund demonstrated that the cylindrical leaf sheath of grass seedlings is consistently more electropositive at its tip than at its base. Such longitudinal electrical gradients are correlated with growth patterns. Recently, Ian Newman, an Australian plant physiologist, showed that these gradients are changed when the tissue is exposed to red or far-red light perceived by the plant pigment phytochrome.

When kept erect and in a uniform environment, leaf sheaths, stems, roots, and other cylindrical plant organs normally show no transverse potentials. If the organ is laid on its side or is exposed to unilateral light, however, it quickly develops a transverse electrical potential. In leaf sheaths, the side that becomes electrically positive will grow more rapidly than the other side. The resultant inequality in growth rate leads to the familiar curvature of the leaf sheath toward light and away from the center of the earth.

Plant physiologist A. R. Schrank, working in Lund's laboratory in Texas, demonstrated that both transverse potentials and growth curvatures induced by light or gravity can be prevented by passing about fifty microamperes of direct current through the tissue. This result depends on the orientation of the electrodes, for if their signs are reversed, the current will augment rather than impede the effects of unidirectional light or gravitational field. The ob-

The Limits of Plant Power

vious implication is that the electrical field induced by light or gravity is central to the curvature response.

All agents that cause the curvature of seedling grass leaf sheaths also cause an unequal lateral distribution of auxin—a plant growth hormone. For many years it was assumed that the various unilateral stimuli worked by first producing the transverse electrical potential. This in turn could cause the auxin to become asymmetrically distributed, leading ultimately to altered growth rates and curvature. But a few years ago, biophysicists Lennart Grahm and C. H. Hertz, working in Lund, Sweden, proved convincingly that the transverse electrical potential could not develop if the plant were deprived of auxin. If, after illumination, auxin was then administered, it became asymmetrically distributed, so that there was more on the "dark" side than on the "light" side. Following this auxin redistribution, the electrical potential appeared.

The mechanism leading to auxin redistribution is still unclear, but the effect of auxin in making possible the appearance of an electrical potential may be related to its recently discovered role in speeding up the activity of a membrane-based hydrogen ion (or proton) pump. Such a pump, like the sodium-potassium pump of nerves, could have a marked effect on the generation and maintenance of electrical potentials.

One of the most appealing of the newer experiments in plant electrophysiology has recently been performed by biologist Lionel Jaffe and his colleagues at Purdue University. The eggs of certain brown seaweeds, like *Fucus* and *Pelvetia*, are perfectly symmetrical spheres. Shortly after fertilization, they develop an asymmetry—one side, the rhizoidal pole, bulges out to form a tube that ultimately becomes the rootlike rhizoid. This structure develops into the bottom, or holdfast end, of the plant, while the opposite end of the egg goes on to form the flat thallus that we recognize as the characteristic brown rockweed. The appearance of this bipolarity in the spherical egg can be controlled by a variety of factors, including light. Unilaterally illuminated fertilized eggs uniformly form rhizoidal poles on the shaded side. Jaffe used this effect to investigate the bioelectrical properties of the system. If eggs were lined up in a row in a narrow tube and not subjected to further influences, they developed rhizoids at random, and no measurable electrical potential existed across the two ends of the tube. If, on the other hand, the cells in the tube were first oriented by unilateral light so that they all developed rhizoidal poles in the same direction, the eggs then acted as if they were part of an electrical series and built up measurable electrical potentials across the ends of the tube. The developed potentials correlated well with both the percent of eggs that germinated and the length of their growing rhizoids.

In a parallel experiment, Jaffe placed a layer of polarized eggs on a perforated metal sheet containing pores whose diameters snugly fit the eggs. This assembly of light-polarized eggs acts like a calcium pump; calcium is taken in more rapidly at one end and extruded more rapidly at the other. It therefore appears that the ion pumping and the developed electrical potential are both closely related to the processes leading to the polarization of growth and to the ultimate differentiation of the plant body. Unlike some of the claims of electrical response of plants to psychic human stimuli, these experiments can be readily repeated by independent and unbiased investigators in laboratories all over the world. The findings of Jaffe and others are certain to bring new insights and vigor to the study of plant bioelectrics. A knowledge of these basic phenomena is a necessary foundation for experiments that purport to demonstrate the response of plants to any external stimulus.

Life, Death,
Immortality,
and Other Problems

15

The Immortal
Carrot

THE HUMAN ANIMAL, whether herbivore or carnivore, is totally dependent upon the photosynthetic activity of green plants for its sustenance. Whether we eat a spinach leaf or a beef steak, we are indirectly taking solar energy captured by the chloroplast of a green leaf and transforming it into chemical-bond energy for maintenance and growth of the human body.

One of the greatest accomplishments of agriculture is the development of high-yielding strains of plants, especially cereals. With recent knowledge of plant nutrition, of growth-regulating chemicals, and of chemical and biological control of noxious fungi and insects, man has achieved ever greater productivity per acre. The venturing of many conglomerate corporations into our highly technologized "agribusiness" shows that fantastically high capital outlays per acre under production can be justified in terms of the present high yields.

Despite these agricultural triumphs and the rosy profit

projections of well-heeled corporations moving into the field, those who understand plants realize the precariousness of agriculture's present high productivity. There is a great need for experimentation in new genetic strains of plants. The corn blight in the United States and the devastating rice wilt in the Philippines show that even our most dependable strains of cereals are constantly endangered by mutating pathogens. We are in a continuous race with the fungi and insects, attempting to produce new types of plants that will yield high quantities of food and resist the ravages of pests, which could consume the crop before man harvests it.

The highest-yielding strains of plants are particularly susceptible to pathogens. Because of their rapid growth and maturation, these plants are extremely succulent and usually lack the protective devices against pests that wild plants normally have. In addition, modern intensive agriculture, with forced rapid growth of the same plant on the same soil year after year, is most conducive to the development of pathogenic organisms in the soil. As a result, modern productive agriculture has become truly dependent on a vast arsenal of chemical pesticides.

It has been the geneticist and plant breeder who have traditionally developed new strains of agriculturally useful plants. Through selection, crossing, and selection again, they have developed plant strains with desirable characteristics. In a great many cases, disease resistance from wild strains has been successfully introduced into high-yielding crops. Geneticists have also taken advantage of chance mutations, and more recently, using radiation and mutagenic chemicals, they have accelerated the creation of new strains. Whether the new line arises by chance or is induced by chemicals or radiation, its characteristics frequently can be introduced into existing plant strains with beneficial results.

The Immortal Carrot

Despite these practices, many experts feel that we are in danger of losing our race with the fungi and other pathogens. For one thing, the breeding cycle of a higher plant encompasses, at the very least, several months and often several years. You cannot see the results of a genetic cross until you have harvested the seed, grown the plant, tested it, and then decided what to do in the next breeding operation. In this period some fungi will have gone through many breeding cycles, and may have produced new types, which in some instances will overcome the resistance originally bred into the crop plant.

We need faster techniques for changing plants. Several recent developments in the field of plant physiology give promise of meeting this need. While none of these procedures has become economically important yet, it seems only a question of time before a new technology flowing from these developments will yield something of major importance for man.

In the mid-1930s, two Frenchmen, Gautheret and Nobécourt, independently discovered that small pieces of tissue cut from a carrot root and put into a nutrient medium would grow and divide, apparently indefinitely. The carrot plant normally lives for only two years, producing the fleshy root the first year and the flower and seed stalk the second. Yet carrot tissues isolated by these two men more than forty years ago are still living in culture in several laboratories around the world. As far as we know, these plants are potentially immortal.

The medium for plant tissue growth must have the usual mineral salts required for the growth of all plant tissues, plus a source of carbon, such as sucrose, for energy and the building up of carbon skeletons.

Only two additional substances are required: these are plant hormones, known respectively as auxins and cytokinins. Although they are present in only infinitesimal quantities in tissue, a balanced supply of these hormones is absolutely required for the normal growth and development of plant cells. It was discovered in the late 1950s that changing the relative concentrations of auxin and cytokinin leads to the development of different growth patterns in cultured tissues. An excess of auxin with respect to cytokinin in the medium, for example, leads to the initiation of roots. Similarly, an excess of cytokinin leads to the development of buds. A ratio of auxin to cytokinin between these two extremes favors the development of undifferentiated plant tissue, called callus.

Knowing these facts, a scientist can take a bit of tissue from almost any plant and grow it in a flask in his laboratory. Furthermore, he can at will regenerate the entire plant by changing the hormone concentrations so that roots and stems form. Using this technique, several investigators have grown entire plants from small explants of tissue. The newly formed plants were normal in all respects, including their capacity to produce viable seeds.

As smaller and smaller bits of tissue were used in the experiments, it was ultimately demonstrated that only one cell is required for the regeneration of an entire organism. This is definite and elegant proof that within the genetic material of each cell there exists the complete blueprint for the development of the entire organism.

On a practical level, the demonstration that a single cell or small groups of cells can be used to propagate many copies of an organism is useful in the creation of plant strains where seed production is a long and complicated process, as for example, with many forest trees. If, from a single young seedling, tissue is excised and grown in culture and then bits of this tissue are made to differentiate roots and buds, numerous new seedlings can be produced with a tremendous saving in time. Some valuable genetic material

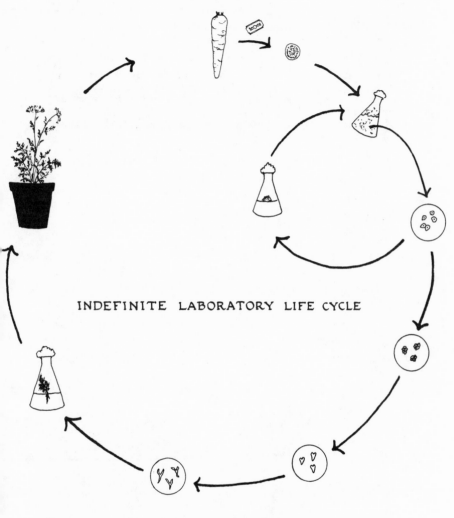

INDEFINITE LABORATORY LIFE CYCLE

In the laboratory, an artificial life cycle can be created for the carrot. Sections taken from the fleshy carrot root can be put into shaking liquid tissue culture, where individual cells are sloughed off. These cells, when transferred to fresh flasks containing semi-solid medium, will produce clumps of tissue. When properly treated, cells may also give rise to *embryoids,* which grow into small plantlets. These produce the mature plant and more fleshy root.

In nature the carrot is a *biennial*. During the first year the plant grows vegetatively, stores a good deal of food in the fleshy root, and receives a cold stimulus over the winter. In the second year, plants which have been adequately cold treated can respond to long days by flowering. The flower produces many seeds, which fall to the ground and recommence the cycle. The plant bearing seed dies.

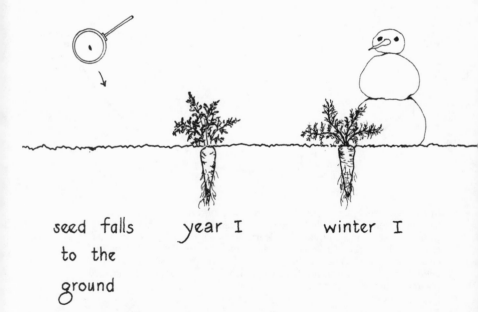

seed falls
to the
ground

year I

winter I

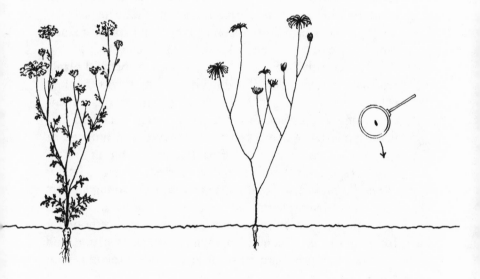

year II	winter II	seed falls
long photoperiod	whole plant	to the
induces flowering	dies	ground

YEAR LIFE CYCLE

has already been introduced into agriculture through this technique, and it is likely that the process will be extended greatly in the future.

A second approach to breeding plant strains also uses the tissue culture technique, but, in effect, avoids one of the results of sex. The cells of the higher green plants, as well as those of higher animals, are predominantly diploid, that is, they contain two complete complements of chromosomes. One set is derived from the maternal side of the cross, one from the paternal. In such diploid cells, a mutation or other genetic transformation in a chromosome of one set is apt to be difficult to detect. For example, a mutation in the maternal chromosome may be screened by the unaltered homologous chromosome of the paternal set. This problem can be overcome by using haploid tissues, which contain only a single set of chromosomes.

It has recently become possible to cultivate haploid plants with comparative ease. To do this, you start with haploid cells. At certain stages in the development of the pollen-producing anthers of flowers, pollen grains or their immature microspore forms may be removed and induced to grow. These are comprised of haploid cells. Thus in certain plants, especially those belonging to the Solanaceae family (including tobacco, potatoes, and tomatoes), young anthers may be cut from the flower, carefully sterilized, and placed in nutrient media. In some cases, the anther will respond by forming tissue, which is haploid. Under stimulation and control by appropriate ratios of auxin and cytokinin, such tissue may produce entire haploid plants.

When such cultured haploid cells are exposed to radiation or mutagenic chemicals, in most cases the mutation can be detected immediately. Tissue of mutated cells usually produces an abnormal product or develops an abnormal nutritional need, such as for a preformed amino acid or vitamin. Thus, the change in the chromosome can be quickly specified, and once such changes are characterized, the tissue may be further treated to yield economically important products.

But most haploid plants are small and weak, and do not yield well. So it is desirable to double the chromosome number, restoring the diploid state. This can be accomplished through the use of the drug colchicine, which is obtained from the crocuslike plant *Colchicum autumnale*. This drug causes polyploidy, in this case diploidy. Under its influence, chromosomes divide but do not separate into two nuclei. Starting, therefore, from a haploid pollen grain, a scientist can detect or induce a genetic change; if this genetic change is desirable, the plant can be diploidized through the drug colchicine, and a new strain of diploid plant obtained rather quickly. At present, not all plants have been successfully treated in this way, and it remains to be seen whether all of them can, in fact, respond to the same techniques.

The creation of new plant strains from a small amount of tissue, from a grain of pollen, or even a single cell may be difficult to imagine, but scientists are already taking another step—they are going inside the cell to build new strains of life. But that subject will be discussed in a later chapter.

16

The Naked Cell

EVERY CELL of every plant lives within a "wooden box," or wall, composed mainly of cellulose, plus a few other less well-defined compounds. Most plant physiologists now believe that the cell wall is a secretion of the living part of the cell, and not a truly integral part of the living system. The individual cells are cemented together by pectins, the materials used commercially in making jellies. Sometimes the integrity of plant tissue is upset through invasion by wall-digesting molds. As these molds make their way through plant tissue, they secrete two types of enzymes: pectinases, which separate cells from each other, and cellulases, which enable the mold to digest the cell walls and to get at the living contents of the cell. If we extract these enzymes from the mold, we can use the pectinases to separate cellular masses into individual cells, and the cellulases to digest the wall away from the living protoplast inside the cell. We have, then, a technique for isolating and unwrapping the protoplast, or most elemental unit of plant life.

The protoplast of a plant cell is an active osmotic system; that is, it will quickly take up water from the surrounding medium and expand. Normally, the protoplast, like the bladder of a basketball, is restrained by the rigid outer coat, the cellulose cell wall. But once that wall has been digested by the cellulase, there is no limit to expansion. The protoplast will absorb more and more water and swell until finally, like an overblown balloon, it bursts. To protect the protoplast against such an untimely end, you simply add osmotically active material, such as mannitol (a sugar alcohol), to the medium around the cell. The protoplast becomes equilibrated and does not swell further.

What can we now do with such naked protoplasts, suitably osmotically stabilized and immersed in an appropriate culture medium? In a significant number of cases, protoplasts will re-form cell walls, and will then go on to divide to produce callus masses, or undifferentiated tissue. These behave in a flask like the ordinary callus masses I mentioned in an earlier chapter. They will form roots in response to high auxin levels and buds in response to high levels of cytokinin. This shows that a single isolated protoplast bears within it all the information required for the regeneration of the entire plant; the cell wall carries none of the essential information. The cells formed from the protoplasts presumably are immortal, like those taken from carrots by Gautheret and Nobécourt more than forty years ago. And, of importance to plant breeders, the naked plant protoplasts are far more amenable to external manipulation than are cells in their cellulose boxes.

The plant protoplasts resemble animal cells in their changeability of form. Animal tissue can also be cultivated —potentially indefinitely—in an artificial medium, although the medium is much more complicated than the one required for plant cells. The animal cell medium must contain some as yet uncharacterized organic materials obtained from immature embryos or from blood serum. In a typical experiment with animal tissue, fibroblast cells are put into a Petri dish containing a culture medium. The cells behave like amoebas, dividing vigorously and migrating over the interface between the glass and the nutrient medium. Unlike experimenters with plant cells, scientists working with animal cells have not yet been able to get the cells to differentiate into organs or the intact organism.

If fibroblasts from two different animals are put into the same Petri dish, they may come into contact and sometimes fuse. To aid this fusion, scientists usually incorporate an inactivated virus into the medium. Although viruses are relatively large particles, they can nonetheless make their way through the usually resistant cell membrane. Proteins on the surface of a virus have an affinity with certain points on the cell membrane. The penetration of the virus through the membrane must involve a local breakdown of membrane structure. Thus, the facilitation of protoplast fusion by inactivated viruses may be due to a local breakdown of the membranes and the spontaneous reformation of membranes across contact surfaces between cells. This could cause cellular fusion. Since the virus is inactive, there are no complications due to virus replication in the cell. Recently, inactivated viruses have been replaced by polyethylene glycol, a long molecule that attaches to the membranes of cells and draws them together.

Using this technique of facilitated fusion of animal cells in culture, experimenters have produced in a Petri dish man–mouse hybrids, chicken–hamster hybrids, and a whole host of other combinations that would be impossible to obtain by conventional techniques. Thus, by such parasexual genetic techniques, you can examine the consequences of introducing foreign sets of genes into host cells.

Unfortunately, in the animal experiments just described, nuclear fusion is usually incomplete. In most cases, one of the genomes is progressively eliminated. Thus in man–mouse hybrids the human component becomes diluted until it almost completely disappears. In other instances, parts of the gene set of, say, a chicken cell are incorporated into a mammalian cell and continue to produce proteins that are characteristic of the chicken and not of the mammal. Because of our inability to induce animal cells to differentiate in culture, these experiments cannot yet be carried further to see the morphological consequences of such fusions.

But let us return to the plant. If a single protoplast can regenerate the entire plant, what might happen if two related protoplasts were fused by this parasexual method? Could one take the fusion product, cause it to develop a cell wall, divide into a callus mass, and then regenerate the entire plant? Would it be a normal plant? Would the genes of the two cells be represented in the finished product? Could viable plants be produced in this way? These questions are now attracting the excited attention of many experimental botanists.

Experimenters have already demonstrated that some plant protoplasts can be made to fuse. The trick appears to be the use, not of inactivated viruses, but simply of an appropriate concentration of polyethylene glycol, or a high concentration of calcium after treatment of the protoplast with alkaline solutions. Fusion is frequently seen, and fusion products can form cell walls. In a few cases they go on to divide, producing undifferentiated cellular masses. Through appropriate hormonal stimulation, cells of such masses can differentiate into entire, normal plantlets. This "somatic hybridization" has been accomplished between different strains of tobacco and different species of petunia, between tomato and potato, and between two different genera in the mustard family.

Plant protoplasts also have the remarkable ability to envelop, and, ultimately, to ingest rather large particles, such as tobacco mosaic virus particles and even little balls of synthetic latex several microns in diameter. When the foreign particle comes up to the membrane of the protoplast, the membrane forms a sheath around it, and eventually folds together into a vesicle, or blister, containing the particle. The vesicle migrates to the interior of the cell and then dissolves, releasing the particle inside the protoplast. This immediately suggests some new possibilities for transforming cells.

DNA (deoxyribonucleic acid) is the molecular basis of heredity. RNA (ribonucleic acid), a smaller molecule, can act as a messenger, carrying information from DNA to the cell. In numerous experiments, investigators have introduced foreign DNA into microbial cells, usually as a virus-transmitted genome, but even as a DNA solution. In some instances, when the host cell with its foreign DNA replicates itself and divides, it also replicates the foreign DNA. This means that new biochemical potentialities have been permanently introduced into the host cell. Since naked protoplasts can ingest viruses, they should be able to ingest functional DNA and RNA and thus become permanently transformed.

Experiments conducted in several laboratories indicate that foreign DNA can be incorporated into host plant cells, leading to permanent transformation of the host.

Thus it seems possible that in the near future we will be able to isolate protoplasts from almost any plant and feed foreign DNA or RNA into these protoplasts. In some cases the DNA or RNA will be incorporated into the host cell, which would then form an altered mass of plant cells and ultimately an altered intact plant. Through the use of this technique we could acquire an entirely new approach to the production of new plant genotypes.

The Naked Cell

With all the difficulties ahead for the experimenter in this new field of parasexual genetics, you might ask, "Why go to all the trouble? Why not simply continue with conventional genetics and produce new types of plants by the old tried and true methods of pollination, fertilization, and harvesting of seeds?" There are many answers to this question, but the most important is that certain sexual combinations are barred to the experimenter simply because of the incompatibility between the pollen tube of one plant and the stigma of another. In general, you cannot achieve sexual fusion of the sperm and egg through pollination unless the plants are very closely related. This means that some desired crosses could not be achieved through conventional techniques, but the crosses might be possible if greatly dissimilar cells fused in the test tube as naked protoplasts. For example, the tomato and the potato are in the same family, Solanaceae, but cannot be crossed sexually because of incompatibility of pollen with the female parts of the flower. They *can* be hybridized somatically by protoplasts. What will the product be like? Will it produce both tomato fruits and potato tubers? Or potato tubers with the ascorbic acid content of a tomato? Or would it be a useless mélange of the two genotypes, producing none of the benefits? It is, of course, impossible to say until the plants mature, but the possibilities are there.

Let us consider another, even more exciting possibility. On a worldwide basis, the greatest limitation to plant growth is the low level of nitrogen in the soil. Nearly all the nitrogen in the biosphere has come from biological nitrogen fixation, that is, the conversion of free atmospheric nitrogen into a compound suitable as plant food. This is carried out either by free-living organisms in the soil, which use decomposed organic matter as energy sources for the nitrogen fixation and growth, or by the symbiotic nitrogen fixers, which live in the nodules of leguminous

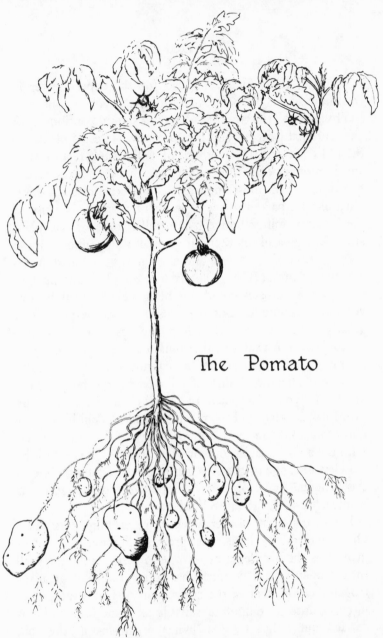

The Pomato

The "pomato." This hypothetical plant can be produced to some extent by somatic hybridization of naked protoplasts isolated from leaves of the tomato and the potato. These two "parent" plants belong to the same family, and the cells are somewhat compatible. The actual "pomato" plants do not produce both tubers and fleshy fruits as shown here. In fact, at present they are rather disappointing and possibly worthless, but the idea is a good one, and future research may produce useful plants by this technique.

plants and certain other species. The occurrence of nitrogen fixation by nodules is so scattered throughout the plant kingdom as to make little phylogenetic sense. If we could take plants that do not now fix nitrogen and fuse their protoplasts with plants that do, then perhaps the nitrogen-fixing nodules could be incorporated into the plants that do not now possess the habit.

What a boon it would be if we could induce corn, wheat, or rice to become nitrogen fixing in this way. Not only would we save the expense of tons and tons of nitrogenous fertilizer, but we would also avoid the pollution caused by fertilizer salts. Best of all, we might greatly increase production of the world's most important class of food plants. At the moment this is a dream, but it might just come true.

17

Molding New Plants

PART OF THE FUN of being a scientist is attending international research meetings. There, one has an opportunity to see and hear colleagues from many countries, and to decide at first hand which of the many reports in the recent literature are apt to provide new insights and start new trends. A recent plant tissue culture meeting in England was especially enjoyable for me. As at all scientific conferences, questions were frequent, searching, and uninhibited. Some extravagant claims, unsupported by adequate data, had to be discounted; some theories, which seemed logical only yesterday, collapsed under the weight of new, previously unpublished data. New theories that emerged from the discussions will undoubtedly be tested in the years ahead and discussed further at scientific meetings. In this way the body of generally accepted scientific knowledge grows and changes from year to year.

A recent advance, described by Albert C. Hildebrandt of the University of Wisconsin, is a new technique for obtaining virus-free plant strains from previously infected ones. When viruses become systemic, they are usually transmitted from generation to generation through seeds. After several generations, virtually every plant will be infected. Sometimes such plants, which look relatively normal, are used commercially, even though the grower is aware that the quality and productivity of his line have declined over the years. Hildebrandt showed how this downward trend could be dramatically reversed in one generation. His method depends on the fact that a virus does not penetrate all parts of the plant body; among the usually virus-free regions are the growing point at the extreme tip of the stem and the anther of the flower. If these organs are cut off the plant, sterilized with alcohol or diluted Clorox, and then placed in sterile nutrient media of the right kind, they will ultimately regenerate entire plants. The new plants will be virus-free and generally much more vigorous than the virus-laden plants from which they were derived.

The stem tip grows first, generating a mass of callus cells, from which roots and new buds eventually arise, producing a new plant. The primordial pollen grains of the anther generally divide a few times and then, as discovered by Colette Nitsch of Gif-sur-Yvette, France, they produce an embryolike structure that also goes on to produce a vigorous, virus-free plant. At the time the pollen cells are formed in the plant, the process of meiosis reduces their chromosome number by half. The plants grown in a culture medium are therefore generally haploid (that is, have cells containing a single set of chromosomes and genes), rather than diploid (having cells containing pairs of chromosomes).

Haploid plant tissues are very useful to the plant breeder. For one thing, they are a source of new, pure-line diploids. All one has to do to develop a diploid is treat the haploid with colchicine, which permits chromosomes to divide but prevents their subsequent separation into two discrete nuclei. Following colchicine treatment, the number of chromosomes per cell will thus be doubled. If the original haploid had six chromosomes (one each of six different chromosomes), the treated haploid, or diploid, will have twelve chromosomes (two each of six different chromosomes). Since each gene on every chromosome will have been exactly copied during chromosome replication, the resultant diploid will be either pure dominant or pure recessive for every genetic character; no gene pair will be present in the hybrid state. Plant breeders usually need about eight generations to achieve an equivalent result.

The haploid plant derived from a pollen grain is also useful—it can be used to produce new types of plants quickly. When there is only one gene of each kind, any mutation in that gene will show up immediately. By contrast, if one gene of a pair in a diploid mutates, the mutation may be recessive and thus hidden by the unaltered homologous normal gene. A haploid plant may be exposed to mutation-inducing chemicals, and mutations of various kinds detected. Most mutations are either harmful to the plant or innocuous, but some are definitely beneficial. For example, plant cells that are resistant to some fungus pest may be discovered after mutation-inducing treatment of susceptible cells. Propagation of the resultant plant by tissue culture techniques may yield a new line of resistant plants.

Certain cells, notably those of bacteria, can have their genetic nature changed not only through breeding and mutation, but also by the introduction of new hereditary material in the form of DNA. The DNA may be introduced by a virus (in a process called *transduction*), or purified DNA may be deliberately applied to the receptor cell (in a process called *transformation*). Can such experimental modifications be accomplished in plant cells? There have been several reports of positive results, but most workers remain somewhat skeptical.

Molding New Plants

Lucien Ledoux of Belgium reported recently that he had corrected a genetic deficiency in a mutant strain of *Arabidopsis*, a small member of the mustard family, simply by flooding the roots of the young seedling with a solution of DNA obtained from a bacterium that did not have the mutant's deficiency. The inference is that bacterial DNA entered the plant, became part of its genetic structure and, by causing the synthesis of appropriate enzymes, corrected the deficiency of the plant cells.

Dieter Hess of Germany similarly reported that when DNA from a red-pigmented *Petunia* was injected into a genetically albino *Petunia* seedling, it caused some red color to appear. Once again, the inference is that the donor DNA entered the receptor and was incorporated into the regular genetic apparatus, leading to the expression of the genetic potential for pigment production. Unhappily, this transformed genetic state did not persist uniformly in the progeny, and Hess's experiment is very difficult to interpret.

Colin Doy and his colleagues in Australia grew a tissue culture of tomato on a synthetic chemical medium. Such cultures cannot normally utilize the sugar lactose for growth, but when treated with a bacterial virus containing the gene for lactose utilization, they were able to do so on a very small scale. C. B. Johnson and his colleagues in Sutton Bonington, England, obtained similar results with cultures of sycamore cells. In both cases, the data are somewhat equivocal because the results are small in magnitude and not clearly genetically persistent. Most scientists remain unconvinced that true transfer of DNA has occurred from the virus to the cell.

What would be accepted as an unambiguous demonstration of the transfer of genetic information from one cell to another? Almost everyone agrees on the format of an ideal experiment. Haploid tissue, derived initially from a culture, should be separated into individual cells through the use of pectinase, an enzyme that digests the intercellular cement binding cells together. These cells should then be placed individually on agar in specially devised synthetic media, permitting them to be tested for various biochemical mutations. Suppose one cell is found that is resistant to a fungal toxin that kills normal cells. Through tissue culture, this new resistant cell should be grown into a plant. It could then be propagated as a haploid or diploid and, in the latter form, might even be useful as a new fungus-resistant strain. The DNA from this resistant plant could be extracted and an effort made to introduce it into the cells of a normal, susceptible plant. The best way to do this is to begin with naked protoplasts of normal cells, prepared by exposing single cells to the action of the enzyme cellulase, which digests cell walls. The membranes of such naked protoplasts can engulf fairly large particles, much as an amoeba surrounds and ingests its food. If the normal protoplast, which is susceptible to the toxin, takes up the DNA extracted from resistant cells, it may become resistant itself. If so, the protoplast should be able to grow on a toxin-containing medium that normally would kill it. If it grows into a plant, and if such a plant behaves genetically like a resistant plant, passing the resistant characteristic on to its offspring in a typical Mendelian fashion, then everyone would agree that the normal protoplast had been transformed by the DNA extract.

Is such an experiment feasible at present? Probably, although only about a dozen genera have been successfully propagated from single, isolated protoplasts. But as the techniques of protoplast culture, cell culture, and DNA isolation are improved, the likelihood of success will increase. I believe that this experiment will be successfully carried out within a decade. Once it is done, all sorts of new tricks will be available to alter and improve the genetic nature of plants. Naked leaf protoplasts from two different species and genera have already been fused to produce a hybrid. It is not too farfetched to believe that new, desirable forms of DNA, extracted from plants and replicated in bacteria, could be introduced into receptor plants through protoplast feeding or through carrier viruses. From such experiments may come higher-yielding crops, proteins of better quality, disease resistance without the use of pesticides, and even an extended range of nitrogen fixation, which would obviate the use of nitrogenous fertilizers. Because of these possibilities, this is an exciting time for botanists.

18

Here Come the Clones

THE NEWSPAPERS are full of stories describing the revolution about to be wrought by genetic engineers and other new breeds of biologists. Who are these people? What are they trained to do? What kind of impact are they likely to have on the world? And should we view their operations with approval or alarm?

We must begin with the realization that man has long been manipulating the genetic constitution of useful plants and animals. Almost all of the strains and varieties utilized in current agricultural practice represent the end points of persistent attempts to improve the breed. The main technique employed so far has been the selection and controlled sexual hybridization of desirable stock. When done judiciously, this can be a powerful procedure. It has, for example, led to the development, from a single ancestral type, of all the current breeds of dogs—from the poodle and Chihuahua to the Great Dane and Saint Bernard. High-yielding hybrid corn and so-called miracle rice and wheat have also been developed by this ancient method. Even the molds that produce penicillin and streptomycin have been improved by selection and controlled mating. Today, a world desperately short of food depends in part on the feverish activities of plant breeders, who strive to produce new strains of wheat and other high-yield crops that will be resistant to wilts, smuts, rusts, and other bacterial and fungal diseases.

In this battle, man is at a serious disadvantage, since those microorganisms that have a rapid life cycle can mutate and produce new pathogenic varieties more quickly than experimenters can select and breed new, resistant types of the slower-developing, high-yielding plants. Hopefully, the newer knowledge of biology, and molecular genetics in particular, may restore the advantage to man.

In animal biology and in fields associated with public health, some of the achievements of recent cell and molecular biology have been noteworthy. Consider, for instance, the technique of cloning, already a reality with various laboratory animals. This procedure for producing many genetically identical individuals depends on the fact that the fertilized egg of most animals, on its way to developing a differentiated embryo, first forms a hollow ball of cells called a blastula. The individual blastula cells, or blastomeres, can be separated from one another without too much injury by the use of selected chemical and mechanical techniques. If this is done early enough in the life history of the embryo, cellular differentiation will not as yet have occurred: each blastomere will be completely "embryonic" and as capable as the fertilized egg itself of producing an entire new organism.

With lower animals, cloning can be accomplished merely by cultivating the blastomeres in an appropriate medium; with mammals, the blastomeres must be implanted individually in the uterus of the natural or a surrogate mother, where they will develop into normal, genetically identical embryos. The latter technique is of great value in rapidly multiplying the numbers of individuals with desirable genetic constitutions. In the dairy industry, for example, it can lead to the swift spread of genes from the most desirable bulls and cows. It has even been suggested that cloning might be applied to humans. We ought to think seriously about this suggestion and attempt to formulate laws to regulate the practice, for while human cloning is not yet technically feasible, there is every reason to believe that it soon will be.

Since almost all of the genetic information encoded in the cell resides in the nuclear DNA, cloning can be accomplished by transplanting naked nuclei instead of entire cells. In such an experiment, the receptor may be either a fertilized egg or an ordinary body cell. First, the nucleus of the receptor cell is destroyed, either mechanically or by controlled radiation, without destroying the capacity of the rest of the cell to continue the various life processes. Next, a nucleus extracted from another cell of a desired genotype is microinjected into the enucleated receptor. Under appropriate conditions, the receptor cell will then develop into a normal embryo, its genetic characters being conferred almost entirely by the DNA of the introduced nucleus. This would obviously be the preferred method of genetic engineering if an egg or other whole cell introduced into a uterus for cloning were rejected by the mother.

But even this means of nuclear transplantation seems crude in comparison with more sophisticated techniques now being developed. One new method involves the isolation from donor cells of individual chromosomes and their introduction into receptor cells. This technique takes advantage of the stage of nuclear division at which the DNA of the nucleus has been condensed into short chromosomes lined up on the equator of the cell. If a cell is broken open at that time, the chromosomes can be separated from the rest of the cell material by gentle centrifugation, in which cell particles are deposited according to their density. Individual chromosomes, however, differ in size and density themselves; by careful centrifugation in a solution of graded density, one chromosome can be separated from another.

When this experiment is done with many cells, controlled so that they enter the crucial phase of nuclear division synchronously, large numbers of identical chromosomes can be obtained. As we begin to learn which chromosomes bear particular genes, we may be able to select at will, for introduction into a receptor cell, those groups of genes whose genetic constitution we wish to vary. In some experimental systems, this basic operation has already been carried out. Here, too, it is just a matter of time before the procedure may be used on human beings.

The genetic engineering techniques described so far involve progressively smaller and more precise intervention into the developmental process. Cloning operates at the cell level, nuclear transplantation involves only that portion of the cell carrying the major part of the genetic material, and chromosomal insertion deals with that part of the nucleus bearing the genes. The ultimate precision of genetic engineering is realized when single genes—or at most, small clusters of genes from a particular chromosome—are multiplied, purified, and then introduced into the cell in which they are supposed to act. In this process, bacteria, whose genetic apparatus is comparatively well understood and amenable to manipulation, are used as a "farm" in which to grow many copies of a desired gene.

Let us follow in detail this particular use of genetic engineering to see how a biochemical lesion caused by a mutation might be corrected. Our example will be a hereditary human disease called galactosemia, in which infants are made seriously or even fatally ill by ingesting milk. Milk contains lactose, or milk sugar, which is split into two simpler sugars, glucose and galactose, during the process of digestion. In the body, galactose is normally converted first to galactose-1-phosphate, then to other compounds. In galactosemic infants, a mutant gene controls the production of an enzyme that is incapable of further transforming the galactose-1-phosphate. The galactose-1-phosphate, accordingly, accumulates, and when it reaches high concentrations, it damages or kills the infant's cells.

The potentially damaging effect of this mutant gene may be circumvented by putting the infant on a diet free of milk and other foods containing galactose. Later on in life, when bodily development is largely complete, restrictions on diet may be relaxed. This regimen, while it avoids the deleterious effects of the disease, does not cure it. A more ambitious technique involves the proliferation in bacteria of the gene controlling normal galactose-1-phosphate metabolism, followed by the introduction of these genes into human cells.

The chromosome of a bacterial cell is a huge circular loop about 1 millimeter in length, which is about 1,000 times the length of the entire bacterial cell. This giant circular chromosome contains some 5,000 genes, each coding for a protein. Inside the bacterial cell there usually are other, smaller loops of DNA that belong either to viruses or to smaller, independent parts of the bacterial genetic material. The small, DNA loops, called *plasmids*, divide independently of, but usually synchronously with, the other loops. Sometimes, during synchronous replication, the DNA strands of the smaller plasmid loops join with the DNA loop of the large bacterial chromosome to produce a form, consisting of one large and one small circle, that resembles a figure 8. Later on, through an event called crossing over, the DNA strands of the two loops will join, creating a single large circle. Another random crossover event may again liberate an independent small circle from the large loop, but since the second crossover may not have occurred at exactly the same site as the first crossover, some bacterial genes may now be present in the smaller plasmid circle. Because bacteria have genes for properly metabolizing galactose-1-phosphate, we may now have a virus plasmid containing the gene for curing the galactosemia lesion. In that case, all we have to do is propagate these unusual viruses until we have many of those viral particles. They can then be introduced into human cells in a tissue culture so chosen that the only cells able to survive will be those that, in fact, have taken up viral particles containing the normal gene. In this way, galactosemic human cells in culture have actually been cured of their biochemical deficiency, benign viruses being used as the means of transferring genes between bacteria and human cells. Curing a galactosemic individual presents some additional technical problems, but these are certainly not insurmountable and will be conquered in time.

An extension of the above technique may lead to other genetically engineered cures, perhaps for diabetes (caused by deficient insulin production), gigantism and dwarfism (caused by aberrant pituitary function), sexual aberrations (caused by abnormal steroid hormone metabolism), and other genetic abnormalities.

There are dangers as well as possible benefits from this new technique. If genes for resistance to penicillin and streptomycin or genes for producing malignant tumors are introduced into bacterial viruses, there is always the risk that the new genetic combinations may give rise to pathogenic organisms insensitive to all known antibiotics or to cancer-forming bacteria. In recognition of these dangers, a group of prominent scientists called several years ago for a moratorium on genetic engineering experiments, at least until a proper series of guidelines can be worked out to control the probable accidents. But even if agreed to by all scientists, such a moratorium would be difficult to enforce, and an unscrupulous group might even use these techniques as a means of fabricating new weapons of war.

As always, advances in scientific knowledge open Pandora's box ever wider. Will humans be sufficiently intelligent and disciplined to handle the dangers? We will soon find out.

19

Plant Cancer

THE ORDERLY DEVELOPMENT of any multicellular organism, whether plant or animal, depends on a system of controls and mutual restraints among the different organs and even the different cells within each organ. How a fertilized egg develops into a complex organism, such as a rabbit or a tomato plant, is still a largely unsolved mystery, involving the complex processes of cellular differentiation. Molecular biologists have recently given us some insights into the events governing the emergence of differences among previously similar cells.

What seems to be involved is a regulated switching-on of some of the genes in the cell; these genes, through the production of specific messengers, then govern the production of specific proteins, including enzymes that catalyze the basic biochemical reactions of the cell. It can easily be imagined that different assemblies of switched-on genes can lead to the production of different populations of enzymes, and thus to cells differing ultimately in form, chemistry, and function. The key to the differentiation process, according to this theory, would be the master mechanism that decides which genes in a particular cell will be switched on.

It has now been shown, in both plants and animals, that most—if not all—cells of an organism have the same total assembly of genes. Thus, a whole carrot plant can be reconstituted from an individual carrot root cell if the cell is cultured in an appropriate medium. Similarly, nuclei from the intestinal cells of a toad, when injected into an egg cell whose own nucleus has been removed, can preside over the formation of a normal animal. This means that cellular differentiation is not irreversible and that cells with certain morphological and chemical characteristics can ultimately assume different characteristics. It all depends on which genes within the cells are switched on, which off.

What keeps a gene switched off? Probably its DNA is being covered by another large molecule, which prevents the gene from forming its messenger RNA. Such large *repressor* molecules are, at least in the bacteria, proteins that have a specific affinity for a particular gene.

Anything that removes the repressor from the gene has the effect of switching it on, and is called an inducer. Among the best-known inducers are the comparatively small molecules known as hormones. In both plants and animals, hormones migrate from the cells in which they are synthesized to some target tissue. There the hormones either stimulate the cells to new biochemical activities or switch off other activities. This system of mobile inducer hormones is one of the most powerful techniques used by plants and animals to insure orderly, balanced development. If cells of part of a plant are dependent on a substance produced by cells of another part, then the development of the two parts must proceed together.

In a plant, a proper balance of roots, stems, and leaves is maintained by a combination of nutritional and hormonal interactions. Roots absorb water and minerals from the soil and send these essential materials to leaves and stems;

conversely, leaves produce photosynthetic materials and send them to stems and roots. This nutritional interdependence guarantees a proper balance between root mass and leaf mass. If a root system is too small, leaves and stems would be deficient in water and minerals. And if a leafy apparatus is too small, the stems and roots would starve. Superimposed on this gross nutritional control is a finer hormonal control; for example, roots produce hormones essential for leaf growth and maintenance, young leaves produce hormones essential for stem and root growth, and cells in the stem tips can influence the growth rate of other stem cells far removed from the tips.

Sometimes this orderly system of controls is disturbed, and the organism develops abnormal growths. Galls grow on oaks and other plants when insects lay eggs there and the larvae develop within the plant tissue. We assume that the insect produces substances that regulate growth and alter the harmonious integration of plant tissues into their normal structures. Other malformations can be produced by bacteria; one causes witches–broom by preventing the normal elongation of stem internodes, another causes clubroot in such plants as cabbage, and a third bacterium can cause the appearance of malignant tumors known as crown galls. This last plant disease is especially interesting because of its resemblance to animal cancer.

To induce a crown gall, one must first wound the plant. Then virulent cells of *Agrobacterium tumefaciens* must be introduced into the wound. After about two to three days, the cells near the wound have been invaded by the bacterium and transformed into a malignant tumor. These tumors continue to grow and, unlike normal plant organs, their growth does not come under the hormonal control of any other tissues. The cells of the crown gall seem to circumvent the normal controls, producing by themselves all the

A witches-broom on hackberry, caused by a bacterium. The normal pattern of branch formation is altered, probably because the bacterium secretes hormonally active substances that "confuse" the plant.

hormones they need for growth. We thus assume that the bacterium in some way switches on previously inactive genes, and that the cancer-like growth occurs because cells gain a new ability to synthesize their own growth hormones.

After some time, the crown gall disease becomes systemic. New tumors grow at other locations on the plant, generally in strained positions such as the angles between the leaves and the stem. These secondary tumors may also grow to very large size, but they differ from the primary tumors in one important respect: they apparently contain no bacteria. Despite the absence of the bacterium that initially caused the disease, these secondary tumors are as malignant as the primary tumors. If either secondary or primary tumors are removed from the diseased plant and grafted onto a normal plant, other tumors will form on the plant. This makes the analogy with animal cancer rather complete; the bacterium is not the true cause of the malignancy, but only the agent that introduces the true infective principle into plant cells.

Why do these cancerous growths occur? What is the Tumor Inducing Principle (TIP)? It turns out to be a plasmid, now known as the tumor-inducing (Ti) plasmid, which contains genetic information that transforms the host cell. Once released into the plant cell, this plasmid may replicate and spread throughout the plant, causing the symptoms of the disease. It appears to cause affected cells in stems to produce large quantities of such hormones as auxins and cytokinins, which stem cells normally obtain from other parts of the plant. When crown gall cells are excised from the plant and grown in pure cultures, they do not require additions of these hormones, as do normal cells. The cancerous cells apparently have many genes switched on by TIP, so that they now have many more capabilities to produce hormones than normal, partially repressed cells. They also have abnormally leaky membranes and lose much material from their cells into the surrounding medium.

The introduction of a cancerous crown gall condition to a healthy plant is neither an all-or-none phenomenon nor is it irreversible. During the crucial two- to three-day induction period after the initial wounding, the onset of the disease may be partly or completely prevented by "pasteurizing" the plant at slightly elevated temperatures. This treatment seems to inactivate the Ti plasmid at a crucial stage of the invasion. Once the disease has set in, it cannot

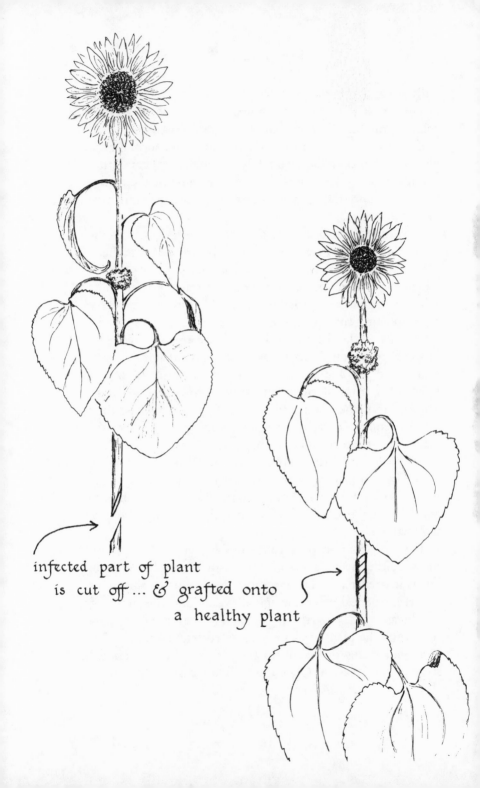

infected part of plant
is cut off ... & grafted onto
a healthy plant

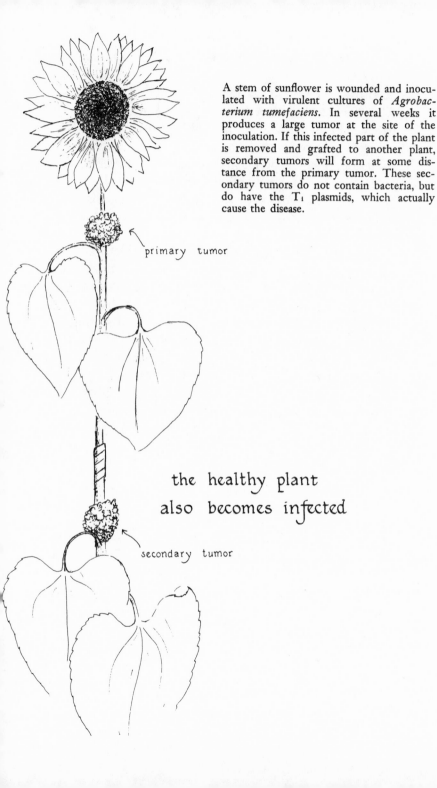

A stem of sunflower is wounded and inoculated with virulent cultures of *Agrobacterium tumefaciens*. In several weeks it produces a large tumor at the site of the inoculation. If this infected part of the plant is removed and grafted to another plant, secondary tumors will form at some distance from the primary tumor. These secondary tumors do not contain bacteria, but do have the T_1 plasmids, which actually cause the disease.

primary tumor

the healthy plant
also becomes infected

secondary tumor

be further prevented by the high temperature treatment. Also, if a weakly induced crown gall is grafted successively to several rapidly growing normal host plants, the gall tissue gradually reverts to a more and more normal growth condition. These experiments indicate that the state of the tissue involves a competition between normal and Ti-induced biochemical events; whichever predominates determines whether the cell will develop a normal or a tumorous growth habit.

This finding has led some researchers to propose that cancer should not be treated by applying growth-retarding chemicals, such as the chemotherapeutic agents now used, but rather growth-promoting chemicals that, by favoring the development of normal components, will swamp out the malignant growth. Whether or not this notion is true, crown gall and other plant tumors are interesting model systems for the study of cancer and offer important insights into the mechanisms that control normal growth and development of all living organisms.

20

In Search of the
Antiaging Cocktail

LIFE has been called a fatal disease. For all but a few organisms, death is an inevitable consequence of existence. From a biological point of view, this makes good sense, for death removes existing organisms and recycles their components, thus providing both environmental niches and basic chemicals for new life. The few exceptions to this almost universal rule include both the smallest and largest of living things.

Bacteria and similar microorganisms seem able to live indefinitely, granted only the availability of food and the removal of waste products. When cultured in this manner, such organisms reproduce by cell division at a constant rate that seems unaffected by the passage of time. One has to conclude that under appropriate conditions, these minuscule forms of life are at least potentially immortal. The only sense in which death occurs is that the "mother" cell disappears in the process of giving rise to the two "daughter" cells. Among the larger forms of life, individual giant redwoods and bristlecone pines seem able to go on living and growing for thousands of years. Possibly they could live forever or at least until disease or some accident, such as lightning or fire, puts an end to their lives.

For most species of animals, there seems to be a built-in timer that dictates the maximum life-span. For humans, the average life-span may be approximately the biblical three score and ten years, although one Quebec man is recorded as having lived to 113. (Stories of 160-year-old residents of the Caucasus and Azerbaijan must be treated skeptically because of the unavailability of birth certificates or other proof of date of birth.) Dogs generally die at the age of ten to twelve years, although several have lived beyond thirty, horses live up to thirty years, elephants live an average of twenty-five years, while houseflies last about seventy days.

As is widely known, advances in medicine have resulted in an increase in the average human length of life, but this is not due to an increase in the upper limit of the life-span. What preventive medicine and public health measures have done is reduce the number of early deaths from accidents and diseases associated with maternity, infancy, and childhood. As a result, the population profile, which used to look like a squat pyramid—with a large population base at birth and a sharp apex at the age of death—now looks more like a column of uniform width from birth until relatively old age, at which time there is a gentle taper to a point. Given contemporary medical advances, there is now no correlation between age and death rate below the age of thirty; beyond that point, the probability of death doubles with each eight years of increasing age. This is generally taken as meaning that the degenerative processes leading to death set in at about the age of thirty.

Within any given species, there is some variation in average life-span, and sometimes these variations can be traced to definite causes. The most obvious cause is the genetic background of the individual. Several decades ago, the American biologist and biometrician Raymond Pearl showed a positive correlation between the age of an individual at death and the age of that person's parents at death. Pearl found that 45.8 percent of those who lived beyond age seventy had parents who did likewise, while only 13.4 percent of the septuagenarians had two short-lived parents. These figures seem to indicate that the best way to live to a ripe old age is to receive genes for longevity. And an obvious inference is that selective breeding among extraordinarily long-lived individuals might lead to an extension of the normal life-span. Unfortunately, children cannot choose their parents nor can individuals who wish to become parents sense the ultimate age of their mates.

Direct experimental proof of the connection between genetics and age at death has been demonstrated by several researchers. Working with rotifers—a simple form of animal—Albert Lansing, a specialist in the physiology of aging at the University of Pittsburgh School of Medicine, was able to modify the creatures' normal life-span of twenty-four days by simple selection. Eggs taken only from adolescent rotifers were grown, and the process was continued for more than fifty generations. The average life-span of the resultant individuals was found to have increased to more than one hundred days. Contrariwise, repeated selection of eggs from senescent rotifers for more than fifty generations decreased the life-span to fourteen days. The generalization seems clear: young mothers give birth to longer-lived offspring; older mothers to shorter-lived offspring.

Robert Sokal, a statistical biologist at the State University of New York at Stony Brook, achieved the same result as Lansing but in a different way. Working with flour beetles, he killed all beetle mothers shortly after they had produced their first eggs. Within forty generations he succeeded in decreasing the average life-span of the resultant population. Sokal reasoned that this change occurred because all those beetle mothers with potentially longer life-spans had had their lives cut short, while those with potentially short life-spans lived to their full life expectancy. The result, in a kind of test-tube evolution study, was a drift of the population toward shorter life-spans.

In some species, including the human, there is a correlation between gender and longevity. Thus, in our society, women live an average of more than five years longer than men. Nobody yet knows whether this is due to the genetic difference between males and females (females have twenty-two pairs of autosomal chromosomes plus a pair of X chromosomes; males have the same number of autosomal chromosomes but substitute one longer Y chromosome for one of the X chromosomes) or whether the earlier death of men is related to a difference in their life styles. Whatever the cause, this differential mortality results in a marked excess of females in the older population.

Other factors affecting life-span in various species are the temperature at which young are reared (particularly in animals whose body temperature is not stable), diet (especially in early development), and exposure to radiation. Fruit flies, accordingly, live ten times longer at a temperature of 10° C. than at 30° C.; rats given a barely adequate caloric diet live longer than their well-fed siblings; and the life-span of mice exposed to whole body irradiation decreases progressively as the radiation dose increases. These experimental data have led certain experimenters to reason that aging is caused by some product whose concentration in the body is increased by high levels of nutrition, high rates of metabolism, and high doses of radiation. It has also led to attempts to ameliorate the harmful effects of these factors by dietary means, which might act as a kind of fountain of youth. For example, radiation is known to cause the formation of free radicals, or molecules with unpaired electrons. Such molecules are extraordinarily reactive, and the harmful effect of radiation may result from the modification of nucleic acid and other important molecules by radiation-induced free radicals. One way to get around this trouble is to feed the body compounds that are known as "reducing agents," or antioxidants. The theory is that such ingested substances will react directly with the free radicals, thus keeping them from affecting important molecules in the cell. Vitamin E, a known antioxidant, has shown some promise in this regard.

One of the paradoxes of the aging process is its alteration when cells are removed from their natural positions in the body and are grown in artificial chemical media. The famous chicken heart culture experiments of Nobel laureate Alexis Carrel, for which Charles Lindbergh designed a pump, showed tissue from that organ could be artificially kept alive for more than thirty years—many times the normal life-span of the chicken from which it came. In fact, that heart tissue might still be alive today if a careless technician had not inadvertently terminated the experiment.

Intimations of immortality are also gleaned from experiments with plant tissue cultures. As I have mentioned, cells removed from a carrot root in 1937 and put in an artificial medium are still dividing and show no sign of running down. In nature, by contrast, a carrot lives a maximum of two years.

Human cells also show evidence of extraordinary longevity when removed from the body. Henrietta Lacks, a woman with cancer of the cervix, provided cells for tissue culture in 1951. The donor has since died, but her cells, the noted HeLa strain used in laboratories all over the world, continue to grow. For such cells, it may be said that mortality is a consequence of being included in a differentiated body. In a sense, experimental immortality is at hand for everyone—simply by donating some skin cells to a tissue culture laboratory, your cells, containing your genes, can be replicated possibly forever.

A different pattern is shown by human fibroblasts—the cells that form the connective tissues of the body. Microbiologist Leonard Hayflick showed that fibroblast cultures were limited to about fifty divisions, accomplished over a period of about eight months. If the cultures were transferred to cold conditions in the middle of the experiment, thereby stopping cell divisions, the cultures would "remember" the number of divisions remaining before they ceased activity. In a companion study, Hayflick showed that the older the individual used as the source of the fibroblasts, the fewer the resulting divisions in culture. These data provided strong support for the "hourglass" theory of aging, in which the buildup of some harmful metabolite (or depletion of some essential substance) is envisaged as the cause of the aging process.

In the human body, which contains nondividing as well as dividing cells, the phenomenon of aging is not likely to be attributable to a single cause. Nerve cells, for example, never divide after their formation. As life progresses, they suffer continual wear and tear until they die. At maturity, the brain has about one billion nerve cells. It is estimated that each day thereafter about 10,000 of them die and are irretrievably lost, never to be replaced. By the age of sixty-five, about 20 percent of all brain neurons are gone; this may account for the diminished mental acuity of some elderly people.

Muscle is another nondividing kind of cell in which wear and tear may be expected to play an important role. Much interest has recently been shown in collagen, the noncellular matrix that binds many cells together, forms the basis for connective tissue, and is the milieu in which bone calcification occurs. The collagen molecules in young persons are highly hydrated and quite flexible; the tissues of the young are accordingly supple. In advancing age, collagen becomes much more highly cross-linked and rigid, possibly as a result of oxidative changes. The resultant rigidity reduces the efficiency of muscles and such organs as the heart and lungs. Since collagen fibers cannot be renewed once they are laid down in the body, the loss of pliability leads to irreversible changes that are part of the total aging process.

But even tissues that can be renewed by cell division show diminished efficiency with increasing age. This may be related to an accumulation of perpetuated "mistakes" in the genetic apparatus. These could be in the form of extra chromosomes or errors at the level of DNA replication that cause mutations. As the load of accumulated errors becomes greater and greater, the rate of replacement of worn-out cells that need renewal (red blood cells, for example) diminishes progressively, and recovery from any injury becomes more difficult. Like the "wonderful one-hoss shay/that was built in such a logical way/it ran for a hundred years to a day" and then suffered total collapse (Oliver Wendell Holmes, *The Deacon's Masterpiece*), the body keeps functioning, all the parts suffering progressive attrition until some breakdown becomes the ultimate insult that terminates life. It is not likely that any single elixir will be able to retard all these degenerative changes. At the very least, the fountain of youth will have to be a complicated, multicomponent, antiaging cocktail.

Offbeat Plants

21

A Living Fossil

IMAGINE the emotions of a paleontologist, long a student of dinosaur evolution, who suddenly encountered a live *Brontosaurus, Triceratops,* or *Tyrannosaurus.* That is roughly what happened about ten years ago to Sanford and Barbara Siegel, University of Hawaii botanists, when they examined the microorganisms of a sample of soil gathered near the wall of Harlech Castle in Wales. When they cultured that soil sample in the presence of concentrated ammonium hydroxide, which greatly inhibits or arrests the life processes of most conventional cells, the medium triggered the growth of microscopic clusters of star-shaped bodies attached to slender stalks. Each body, about 5 micrometers (0.0002 inch) in diameter, closely resembled pictures the Siegels had seen of a recently discovered fossil microorganism. But as far as they knew, no living specimens of this organism had ever been described. With the help of the fossil's discoverer, Elso Barghoorn of Harvard University, the Siegels were able to establish, in a strange sequence of paleobotanical events, that they had found a living relative of an organism first described as a fossil.

Barghoorn had made his own discovery while gathering

specimens of ancient rocks in a search for primitive organisms. One specimen of chert, or flintlike rock, from Kakabek in Ontario, Canada, contained peculiar umbrella-like forms that seemed regular enough in physical appearance and structure to be considered microorganisms, rather than a pattern that had developed as the rock formed. Barghoorn named these microorganisms *Kakabekia umbellata*, meaning umbrellalike form from Kakabek.

Since the rocks in which Barghoorn's forms appeared dated from the middle Precambrian period, about two billion years ago, the microorganisms were among the oldest of all plantlike fossils. The Siegels' discovery of a living relative of Barghoorn's fossils established a remarkable thread of biological history. The Siegels named their creature *Kakabekia barghoorniana* in honor of their colleague.

Kakabekia barghoorniana

ca. 4000 X

A highly enlarged picture of *Kakabekia*. An umbrella-like structure is attached to a slender strand. This creature chooses to live in bizarre environments.

The Siegels had come to Harlech Castle in the course of a long project. For many years, with the support of the National Aeronautics and Space Administration, they had been examining the physiology of organisms under stress, especially stress caused by harsh environments likely to be encountered during space travel or near or on other planets. Because ammonia is generally thought to have been one of the more abundant components of the primitive earth's atmosphere, and is still one of the major components of the present atmosphere of Jupiter, the Siegels wanted to study earthly environments with abundant ammonia. Natural candidates for examination were soils saturated with urine. On a chance visit to Harlech Castle, they observed tourists urinating near the castle walls, and learning that this was an old practice, decided to collect their first soil sample. Since the organism they discovered in that soil thrived in the presence of concentrated ammonium hydroxide and was not seen in other environments, they hypothesized that they had found an obligate ammonophile, an organism that requires ammonia in order to grow.

In the ten years since they first visited Harlech Castle, the Siegels have found that they were mistaken about *Kakabekia*'s need for ammonia. At sea level, the microorganism does require the compound, yet *Kakabekia* also appears in certain mountainous regions low in ammonia but high in alkalinity. Without ammonia, *Kakabekia* needs special soil conditions, as well as peculiar combinations of temperature and altitude, to thrive.

In soil samples from Hawaii, California, the Great Plains states, Illinois, New York, western Europe, and northern South America, the Siegels found no signs of *Kakabekia*. But the microorganism did show up in soils from Alaska, Iceland, and various alpine regions. So temperature seemed to be one of the factors governing the distribution of modern *Kakabekia*. But in studying its distribution up Hawaiian and Japanese mountain peaks, The Siegels found that it grew in bands, which meant that at least one other factor was interacting with temperature to limit the organism's distribution. At lower elevations, *Kakabekia* is not found below about 45° north latitude, and thus appears restricted to regions with comparatively cool summers. At latitudes closer to the equator, it grows only at altitudes above

6,500 feet. Since air temperature decreases about 7° C (about 12° F) for each 3,300 feet of altitude, *Kakabekia* seemed to require low temperature and a certain altitude.

The *Kakabekia* found on Mauna Kea, an extinct volcano in Hawaii, appeared to bear out their ideas about the microorganism's temperature and altitude requirements. They recovered it on the volcano at altitudes above 11,500 feet. Once cultured, organisms consistently appeared in the soil samples at −7° C, but not at 30° C. But at about 10,000 feet up Mauna Kea, and on down to the base of the mountain (about 7,500 feet above sea level), *Kakabekia* showed exactly opposite characteristics, growing at 30° C, but not at −7° C. This second *Kakabekia*, which has adjusted ecologically to warm temperatures, seems to be a variant of the cryophilic, or low-temperature-favoring, strain that the Siegels had found earlier.

There are other variants of *Kakabekia*, with differences in the umbrellalike cap, with or without a stipe, or stalk. The umbrella may be lobed, scalloped, fringed, or cut into rather acute angles, and may have varying numbers of rays, the umbrella's ribs. These strains probably represent genetic variants that evolved from some prototypical form.

Another peculiarity of *Kakabekia* is that its cultures require no oxygen, but unlike typical anaerobic bacteria such as *Clostridium*, which causes gangrene, it is not killed by oxygen in the air. *Kakabekia* does require a distinctly alkaline environment, which can be furnished equally well by sodium hydroxide, potassium hydroxide, or ammonium hydroxide at a level that would be toxic to most living creatures. When sodium hydroxide is replaced by sodium metasilicate, which provides the same kind of alkaline environment, as well as silica, *Kakabekia* grows more slowly, or stops altogether. Since it contains large quantities of silica (the hard part of its cell walls that leaves a fossil) and has almost always been found with diatoms (algae that have silica cell walls), researchers thought that *Kakabekia* could easily take advantage of silicon in its microenvironment. But if metasilicate is toxic to *Kakabekia*, it must obtain its silicon from ordinary minerals in soil, possibly from sand, which is silicon dioxide.

So far we know little about the *Kakabekia* cell and its metabolism. Our information has been limited because we have been unable to grow *Kakabekia* in pure culture—all cultures have been contaminated by other organisms, invariably diatoms—and because *Kakabekia* cells seem to grow very slowly. We do know that when these cells are tested with the Feulgen stain, which shows up the DNA of chromosomes, the results are negative. Since all known living things must contain DNA, *Kakabekia* is probably a prokaryote, an organism that, like a bacterium, does not have its chromosomes gathered into a discrete nucleus. This hypothesis is supported by the discovery by several Russian workers of some star-shaped bacteria that develop in certain kinds of soil or in creek water rich in organic material. Although *Kakabekia* is nearly twice as large as these bacteria, they do have a marked morphological resemblance. Their possible genetic relation is seconded by the fact that all of the star-shaped bacteria have been collected from cool regions.

Kakabekia's slow growth raises questions in itself. In order to grow, it must have oxidizing enzymes that will mobilize energy from its environment. Yet preliminary experiments have revealed that *Kakabekia* contains neither the heme enzyme nor the phenol-oxidizing enzymes that are most organisms' conventional means of producing energy.

At present, researchers' questions about *Kakabekia* far outnumber the answers. The Siegels and other workers will be looking for the microorganism for some time in such scattered places as Point Barrow, Alaska; Surtsey Island off Iceland; the top of Mauna Kea in Hawaii; and Harlech Castle in Wales. When all of the living fossil's long history is clear, microbiologists and evolutionists will have new stories of their own to tell us.

22

Guayule Bounces Back

IN 1910, about 10 percent of the world's rubber and half the rubber used in the United States were supplied by guayule, a small shrub native to the Sonoran and Chihuahuan deserts of Mexico and the southwestern United States. A member of the daisy family, guayule was familiar to the Aztecs. They made rubber balls for a basketball-like game by chewing guayule twigs, spitting out the pulp, and saving the remaining elastic mass of rubber. Early commercial rubber production was simply a factory adaptation of this Aztec technique. The harvested wild plants, containing rubber in individual stem and root cells, rather than in tappable latex ducts, were first dumped into boiling water to remove the leaves (which contain no rubber), to wash the plant, and to coagulate the rubber. The washed and defoliated plant was then ground and the slurry deposited in a flotation vat of hot water containing alkali, which facilitates rubber extraction. The rubber floats to the surface in wormlike strands that can be skimmed, while the rest of the plant, the bagasse, becomes waterlogged and sinks. After this process is repeated several times, the rubber produced is reasonably free of gross impurities such as soil, leaves, and wood. About 20 percent of the rubber's weight will be resins, which make the rubber tacky and cause it to age and crack more rapidly. Soluble in warm acetone, these resins are easily extracted from the rubber, which then is indistinguishable from the product of the rubber tree, *Hevea brasiliensis*.

Guayule

& the cactus wren

A plant of guayule. A cactus wren gives a clue to its size.

Guayule Bounces Back

Hevea rubber production has risen to about five million tons per year, constituting about one-third of the world's current rubber needs. The remaining two-thirds will come largely from the petroleum-based synthetic rubber industry. The Soviets still produce some rubber from the Russian dandelion, but the other 2,000 or so species known to contain rubber are not now exploited commercially. But the rise in rubber use, a projected shortage, and the cost or scarcity of petroleum have meant that guayule and other rubber-bearing plants might be pressed back into economic service. This possibility would be especially useful to the United States, which currently uses one-fifth of the world's rubber supply. In 1974, we imported more than 700,000 tons of natural rubber at a cost of half a billion dollars, making rubber our third largest import of inedible crude material.

I first encountered guayule in 1943 when, as a new Ph.D., I went to work on the wartime Emergency Rubber Project at the California Institute of Technology. As a country at war, the United States needed rubber tires for airplanes, trucks, Jeeps, and cars. But because the Japanese had overrun the Malay Peninsula, 90 percent of the world's rubber was out of our reach. To solve this problem, two projects were quickly organized: one was to produce natural rubber from guayule; the second, synthetic rubber from petroleum. Between 1942 and 1946, during the guayule project, about 1,000 researchers planted more than a billion seedlings over 30,000 acres, producing about three million pounds of rubber. Late in 1945 production from two California factories approximated fifteen tons of rubber per day, while in Mexico roughly four factories turned out twice as much. After only three and a half years, the project scientists knew more about guayule than almost any other plant; had the program continued, improvements in guayule genetics, agronomy, pest control, milling, and purification would have increased production much more. But outpaced by the production of synthetic rubber, the guayule project was abandoned in 1946 and its fields plowed under or burned. Recently, after taking a long look at America's needs for the future, a National Academy of Sciences panel has recommended that the production of guayule rubber be resumed.

Guayule's great virtue is that it is a completely renewable resource, whose use would free for other purposes petroleum now used to produce synthetic rubber. Guayule can be grown in the United States and Mexico, eliminating our dependence on most foreign sources.

Unlike production of synthetic rubber, producing guayule rubber does not cause pollution. Guayule grows well in semiarid regions and needs only about sixteen inches of rain a year, although optimal rubber production does require some irrigation water. And since much of the land on which it could be grown lies on Indian reservations, it could possibly become an economic boon to a generally impoverished segment of the American population.

Early in the twentieth century, a group of financiers, including Rockefeller, Guggenheim, Aldrich, and Baruch, invested $30 million in the Continental-Mexican Rubber Company. This American-financed venture produced much revenue for Mexico, and desert land prices in both the United States and Mexico boomed in anticipation of a continued bonanza. But overuse of wild plants and failure to replant (sixteen million pounds of rubber were imported into the United States in 1912 alone) led to a virtual disappearance of the new industry's basis, and mills were forced to close. In Mexico, the revolution gave the industry its *coup de grâce* as a large operation. Minor operations continued in California. In the late 1920s, Britain's control of Malaya and its resultant rubber monopoly suddenly increased rubber prices threefold (much as the OPEC nations have recently done with petroleum, and Brazil with coffee). In Mexico and California, guayule rubber was profitable again and production resumed. A then Maj. Dwight D. Eisenhower, assigned to study guayule in connection with national security, recommended further development of the industry, but the world depression of the 1930s postponed the project until its short-lived resurrection in 1942.

Mexico, however, has never stopped producing guayule, and at present a pilot plant in Saltillo, in northeast Mexico's Sonoran desert, can process one ton of shrubs daily. About 2.6 million tons of wild guayule grow in an area of approximately 10 million acres in the states of Durango, Coahuila, Zacatecas, Chihuahua, Nuevo León, and San Luis Potosí, just south of the Big Bend area of Texas. The project's goal is 30,000 tons of rubber per year, from a harvest of about a third of a million tons of plant. Hence, the Mexicans have a nine-year supply of guayule: the harvest of about one-ninth of the total number of plants per year would allow adequate regrowth of seedlings. A harvest of wild shrubs obviously depends on a large supply of inexpensive hand labor, available in Mexico, but not in the United States.

If guayule growing were revived in the United States, the main location would probably be about five million acres of the arid zones of Texas and California, connected by a narrow band composed of part of southern Arizona and a small portion of southwestern New Mexico. But as the National Academy of Sciences panel points out, the industry would be economically viable only if it were to benefit from the same kind of research that has led to a tenfold increase in *Hevea* production over the last thirty years. The Emergency Rubber Project showed that guayule can be genetically improved through conventional selection and breeding techniques. Innovations in chemical weed control, insect control, and rubber technology, as well as new insights into the biochemistry of rubber formation in plants, can be expected to increase yields. Agronomic tricks, such as harvesting only the tops and leaving the roots to resprout, might shorten the current three- to four-year cycle before rubber harvest is optimal.

Along with each ton of purified rubber, the guayule plant produces about two tons of bagasse (mainly crushed stems), one ton of leaves, and half a ton of extracted resins. Each of these could find commercial use as a by-product. The bagasse can be used for paper, pressed board, and cardboard manufacture. Guayule leaves are covered by a hard cuticle wax, whose melting point, 169° F, one of the highest ever recorded for a natural wax, makes guayule wax competitive with carnauba wax, most commonly used for polishes. The resins, containing odoriferous terpenes, shellaclike components, cinnamic acid, and drying oils, could probably find some use. Minute guayule seeds could be sources of protein and fat, and the leaves might serve as an occasional browse for sheep and goats. But whatever the contribution of its byproducts, guayule must stand or fall on the quality and quantity of its rubber.

Even if petroleum were indefinitely and cheaply available, demand for natural rubber would grow. Rubber is a polymer, a very large molecule made up of repeating units of isoprene—a small molecule made up of four carbon atoms in a row and one branching off the chain, together with eight hydrogens. When isoprene is polymerized, or linked in a specific way to make rubber, the units join end to end along the unbranched four-carbon chain. The isoprene unit has two double bonds involving four of the five carbon atoms; each of these two double bonds is rigid, whereas the carbons are able to rotate freely around the single bond. This means that there are two possible ways, known to the organic chemist as *cis-* and *trans-* configurations, for the carbon atoms to be grouped around the double bond. In natural rubber from *Hevea* and guayule, the bonds are all *cis*, that is, the branched carbon atoms all protrude from the same side of the backbone of the polymer. This configuration gives maximum elasticity and bounce to the molecule. Even a small percentage of *trans-*links can markedly reduce the desirable qualities of the molecule. In synthetic rubber manufacture, both *cis-* and *trans-*linkages are produced, so that the resultant molecule is weaker than natural rubber. While perfectly acceptable for many commercial uses, the synthetic product does not stand up well in tire sidewalls; accordingly, all tires must have at least some of the natural polymer, both to give basic strength and durability and to serve as a kind of template around which the synthetic units are deposited in a regular way. Since tires account for about 75 percent of all rubber used in the United States, the reason for continued reliance on natural rubber is clear.

The amount of natural rubber required depends on the type of tire. Airplane tires, which must withstand the tremendous, jarring shocks of landings, are composed almost entirely of natural rubber; truck and bus tires, which operate under great stress for prolonged periods of time, have about 40 percent natural rubber; and ordinary automobile tires generally have about 20 percent. (A post-World War II government-sponsored test of a set of commercial truck tires made of guayule found that they performed as well as *Hevea*.) Radial tires have somewhat altered this picture: radials must contain almost double the amount of natural rubber. Since 1972, radial tires have captured more than half the tire market, and this trend will probably continue, underscoring the need for more natural rubber. Since *Hevea* production, already strained, cannot expand indefinitely, guayule looks like the answer. Even with its modest 1940s production level of about 500 pounds of rubber per acre per year over a three- to five-year period, guayule would be economically workable, but with expected improvements in productivity, it may come to match *Hevea*, which consistently produces about two and a half times more rubber.

To the plant physiologist, guayule and other rubber-producing species pose the additional lure of research. Why does the plant make rubber, since rubber is inert metabolically and cannot be used as a food source? How does the plant manage to achieve 100 percent *cis*-bonding as it fabricates the long-chain isoprene polymer? Why is rubber made only in stems and roots, not in leaves? How can guayule compete so effectively with other plants in the arid desert environment? Part of the answer to the last question involves the secretion from guayule roots of a substance that retards the growth of nearby seedlings. This kind of "chemical warfare" is common in desert plants, and in 1944, working with James Bonner at CalTech, I was able to identify *trans*-cinnamic acid as the main effective compound. Oddly enough, this same molecule later found use as an antagonist of auxin, one of the major plant growth hormones. As far as I know, guayule is the only plant known to use a natural hormone antagonist in this way. No doubt this is not the last surprise we can expect from guayule.

Plants and
the Environment

23

How Safe Should Safe Be?

MODERN industry, agriculture, and medicine float on a sea of synthetic chemical compounds. Every year thousands of such new products are devised. Each purports to solve some human problem or satisfy some human need better than its predecessors. Some are uneconomic to make and never reach the production line or sales counter. Others are weeded out between the testing laboratory and the production line because they are obviously dangerous or toxic to human life. But even those that get through the screens imposed by private companies and the various local, state, and federal agencies cannot automatically be considered safe. In fact, an alarming number of compounds and processes, long accepted and used, have recently been found to have unexpected and deleterious effects on biological systems. Thus it has become imperative to inquire closely into the criteria that are, and ought to be, employed to safeguard the public health and well-being: continually to explore the question, How safe should safe be?

Developing these criteria is not entirely an exercise in rational, dispassionate analysis. More and more, the process involves reconciling the often conflicting interests of business, agriculture, and the environmentalists. Known benefits are carefully weighed against demonstrated or possible side effects. The final choices are both subjective and evaluative. DDT is an example. The fact that it can wipe out malaria-bearing mosquitoes must be balanced against its inadvertent destruction of useful insects, such as bees and others serving as sources of food for birds. Similarly, the drop in crop productivity and loss of income that result from the banning of DDT must be balanced against the possibility that its slow biodegradability may ultimately produce new dangers to man. There are still unanswered questions concerning DDT, but while they are being worked out, countries where insect-borne human diseases are still a major problem cannot be expected to ban the compound.

Against this background, a recent discovery by two brand-new Ph.D.s is of particular interest, for by applying a known but neglected approach to the testing of herbicides, they have raised doubts about the alleged safety of most agricultural chemicals in major use today. Michael J. Plewa of the University of Illinois and James M. Gentile of Hope College have produced evidence that atrazine, the most widely used herbicide in cornfields, gives rise to metabolic products that cause mutations, and possibly cancer, in laboratory animals. Independent substantiation of their claims could lead to a massive reappraisal of the procedures normally employed for certifying as safe those chemicals designed to be used in agriculture.

How could such a pernicious effect have been overlooked when atrazine was first tested? Atrazine, itself a Swiss-produced compound, had a clean bill of health. When fed to experimental animals for detection of toxicity symptoms, to microorganisms for detection of mutagenicity, and to tissue cultures for detection of possible carcinogenicity (by induction of cancerous overgrowths), atrazine was innocuous. If it is first supplied to corn plants, however, chemical extracts of the leaves and kernels of such plants show mutagenic activity in appropriate biological test systems. Corn plants not treated with the atrazine do not produce such symptoms. The inference is clear: although itself innocuous, atrazine is transformed by the corn plant into a substance that can cause genetic aberrations. While it has long been recognized that the metabolic products of herbicides, as well as the herbicides themselves, should be tested for toxicity in various organisms, this procedure has not been conscientiously followed with most major compounds.

The results found by Plewa and Gentile have been published so far only in brief and preliminary form. In the meantime, both the federal Environmental Protection Agency and the National Institute of Environmental Health Sciences, part of the National Institutes of Health, have manifested considerable interest in funding a continuation of this study. What makes it so convincing to experts in the field is that these researchers have devised a procedure for detecting possible herbicide mutagenicity within the crop plant itself as well as in microbial test organisms.

To test the mutagenicity of the herbicide directly on the crop plant, Plewa and Gentile used a genetically pure *waxy* corn plant, itself a mutation from standard corn. The gene for waxiness also inhibits the production in the plants of the starch component amylose, made by nonmutant corn. *Waxy* corn produces instead only a related material, amylopectin. Amylose stains a deep blue when exposed to a mixture of iodine and potassium iodide, but amylopectin stains a faint tan. This characteristically different color response to the same reagent applies even in the pollen grains of *waxy* corn, which, because of their haploid chromosomal composition, show up any mutation immediately. This stain reaction accordingly affords a convenient test for the detection of increased mutation rates.

The test is run in the following manner. *Waxy* corn plants are grown in a field without herbicides, and the tassels are collected at flowering time and preserved in a 70 percent ethyl alcohol solution until analysis. At that stage, the pollen grains are removed from the tassels and placed on a microscope slide. The iodine-potassium iodide reagent is added, and in most runs, the reagent turns all the pollen grains tan. Occasionally, however, a pollen grain will stain a deep blue, indicating a mutation from the *waxy* gene back to standard corn. Such a change is produced, it is assumed, by a random mutational event, possibly initiated by a cosmic ray or chemical mutagen in the environment. Whatever the cause, these occasional spontaneous back mutations in the *waxy* gene are found to occur only once in about 100,000 pollen grains. But in plants exposed to as little as 10 parts per million of the atrazine herbicide in the soil, the mutation rate of *waxy* genes is increased to about 25 to 30 occurrences per 100,000 grains. Thus, it appears that atrazine exerts a mutagenic effect on corn pollen when the plant is grown in soil containing even traces of the chemical.

Experimentalists had previously applied atrazine to similarly "labeled" microorganisms, containing genes whose mutation could be easily diagnosed by simple color or growth reactions. Although some researchers obtained positive results, the great bulk of the evidence was negative, and it was on that basis that atrazine had been given its clean bill of health.

How Safe Should Safe Be?

Struck by their strongly positive results on corn, Plewa and Gentile decided to isolate the active material in the atrazine metabolic product for further testing. They ground up the leaves and kernels of their atrazine-treated plants in water, centrifuged away the debris, and kept the remaining fluid. To preserve the extract, they freeze-dried it under high vacuum to a powder. Small samples of the leaf and kernel powder could then be applied in appropriate solutions to the usual microbial test organisms. These include certain yeasts that have found wide use in the diagnosis of mutations and a bacterium that has recently been used to detect mutagenic chemicals in some cosmetics and hair treatment preparations.

In one yeast assay, a mutant—which had originally been produced by a known mutagen—was caused by the unknown atrazine metabolite to back mutate to the standard form. It appeared that the yeast DNA had been converted back to the normal form. Neither pure atrazine itself nor extracts of corn not treated with atrazine produced these effects; the mutagenic activity was thus clearly the result of an interaction between the plant and the herbicide. Similar, although less striking, data were obtained when the same tests were run on two other herbicidal compounds that are related to each other but not to atrazine. These results point to the desirability, even the necessity, of proceeding with equivalent investigations on still other major herbicidal chemicals.

Plewa and Gentile have continued their analysis of the active material in the atrazine metabolite. On thin layers of silica gel, the components of the corn plant extract can be separated so as to yield at least two active mutagens that work on test yeasts and the above-mentioned bacterium. These mutagens are water soluble and probably act by causing a base-pair substitution in the DNA chain making up the hereditary material of the test organisms. The resemblance between atrazine and the four bases that normally make up DNA had previously been noted by other investigators. Even more suggestive in this connection are other herbicides built of substances that are actually modifications of one of the DNA bases. Subjected to the Plewa-Gentile type of analysis, and in light of the experience with atrazine, these herbicides might also be expected to show mutational activity.

This work does not, of course, prove that atrazine-treated corn causes mutations in humans. For one thing, the active metabolic product might be broken down by the acidic conditions of the human stomach or might never be absorbed from the gastrointestinal tract into body tissue. Even if the active material were to enter the body, it might readily be detoxified by the liver or some other body decontamination center. It is also possible that, despite its effect on corn and microorganisms, the substance might not act on humans or animals at all. That, however, would be unexpected, since DNA is similar in all living organisms, and what affects the DNA of one creature should affect that of all.

It appears likely that we will see a marked extension of the kind of testing initiated by Plewa and Gentile. The results may put pressure on the Environmental Protection Agency, the Food and Drug Administration, and the National Institutes of Health to take a position on the continued use of atrazine and related compounds. In the meantime, the organic farming aficionados, who grow only products produced without herbicides or pesticides, would appear to be taking the most prudent course, at least from the point of view of public health.

While no sensible person would claim that we should stop using all chemical compounds in agriculture, mounting evidence indicates that we have not been sufficiently careful in screening these agents before their widespread production and extensive use. Through the serious participation of industry and government, as well as the technical and environmental sciences, it should be possible to find a balanced approach in developing new testing criteria and a better answer to "how safe should safe be?" More stringent criteria may well lead to screening programs that are more complicated, more expensive, and longer than those now in practice. Under these circumstances, alternatives to the use of herbicides and pesticides may become more feasible in terms of economics as well as public health.

24

The Organic
Gardener and
Anti-intellectualism

SEVERAL YEARS AGO, in response to the then strong
student pressures for "relevance" in education and moved
by a desire to consider certain social problems in the frame-
work of the newer knowledge of biology, several colleagues and I initiated a course entitled "Biology and Human Affairs." Designed for nonmajors and actually for
nonscientists (it inevitably became known as "Biology for
Poets"), the course has attracted relatively large numbers
of students and has met with moderate success in its effort
to provide some biological background against which one
can consider such topics as population growth, pollution,
genetics and intelligence, organ transplantation, biological
engineering, and chemical and biological warfare. Teach-
ing such a course is a challenge because of the great diversity of student backgrounds in biology (How much

basic science should I teach?) and because of the inadequate grounding in social science of most biologists, compared with the occasionally great sophistication of some students (Am I out of my depth?). These difficulties have resulted in a constant modification of the course, both in subject matter and approach, a situation that is likely to continue for some years.

Interacting with the students in informal discussions has made several instructors in the course aware of the extent to which many students have become disillusioned with science—as a method of arriving at an understanding of man and the universe in which he lives and as a means of improving the quality of life. To many of the "poets" in the course, scientists are other-worldly people, content to fiddle in their laboratories while the world—for which they are at least partly responsible—burns in napalm, decays into pollution, and becomes dehumanized into mechanical, computerized, assembly-line work routines.

The scientist is also held responsible for the antisocial uses to which many of his discoveries are put by the military (bombs, chemical and biological weapons, flamethrowers, automated battlefields), by government (electronic "bugging" devices, computerized data banks for political dissidents), and industry (shoddy mass-produced automobiles, chemical products that pollute the environment, useless and expensive gadgetry). Many students seem unwilling to agree that the findings of the scientist are ethically neutral, and that it is society that must determine whether they shall be used for good or evil. Because of his special insight and knowledge, the scientist is expected to give guidance to the decision-making agencies of government and business (which frequently do not respect his testimony) and to maintain constant surveillance over the new technology his research discoveries have spawned. Since, with relatively minor exceptions, scientists have not performed this watchdog role adequately, both society in general and students in particular have recently given much less approval to the aspirations and accomplishments of science than heretofore. In addition, some students have reacted with cynical disbelief to some of the data of science, which they would not have questioned several years ago. Let me illustrate with one case history.

As a reaction to the recent exposés concerning the deleterious effects of various pesticides on animal life and the ecosystem, the possible damage to babies from inordinately high quantities of nitrates in their foodstuffs—a result of the application of high quantities of chemical fertilizers—and the known toxic effects of certain common chemicals used as food preservatives, some people, including many students, have become devotees of "organic gardening." Produce grown by organic gardeners is ostensibly never subjected to potentially toxic pesticidal sprays, is fertilized only by manure, leaf mold, or other organic materials, and is never artificially preserved by chemicals. Materials so produced are often more expensive than the usual commercially available foods and frequently must be purchased in special stores.

In the course of my lectures on botany, I had occasion to point out that the green plant is a complete autotroph for organic materials, that is, given only carbon dioxide from air, sunlight, and water, plus inorganic minerals from the soil, it is capable of synthesizing the thousands of organic compounds out of which it is constructed. Under favorable soil conditions, it has absolutely no need for any external sources of organic compounds; and the harvest produced from crops supplied with purely inorganic fertilizers will be just as rich in vitamins, amino acids, and other key nutritional requirements for man as one produced from heavily manured crops. This has been demonstrated many times, not only by analyzing for such components in the laboratory, but also by feeding test animals with both kinds of products. In the absence of evidence to the contrary, almost all scientists would conclude that there is no demonstrable difference, either chemical or nutritional, between inorganically and organically grown plants.

Why, then, the students want to know, is it beneficial to apply manure, leaf mold, peat moss, compost, and other organic materials to the soil? The answer lies, not in the plant, but in the soil itself. Soil is comprised of various-sized particles of degraded rock wetted by a solution containing organic and inorganic materials, together with organic remains of various kinds of creatures, large numbers of living bacteria, fungi and algae, occasional small animals, and a system of air spaces permeating the entire mass. This elaborate mixture of components is not static; on the contrary, it is constantly undergoing transformation. As the organisms grow, the available nutrients in the soil solution and the organic remains are used up. In compensation, these organisms secrete materials that solubilize the rock particles, making new minerals available for plants, and also glue soil particles together into large crumbs, keeping the soil well aerated and in good "tilth" for plant growth.

If the organic matter of the soil is depleted, then the activity of soil microorganisms diminishes and soil quality may deteriorate. Soils poor in organic matter characteristically have poor water-holding capacity, deficient mineral nutrients in the soil solution, and a compacted texture so deficient in air spaces that roots cannot respire properly. Plants growing in such a soil may be mineral deficient for any one of these reasons, and it is to be expected that the addition of chemical fertilizers would only partially alleviate their mineral deficiencies, while manure or other organic additives would do a more complete job.

If, however, the soil has adequate organic matter to sustain its population of microorganisms and to maintain its proper structure, then the addition of chemical fertilizers is sufficient to produce optimal growth and chemical composition of the plants growing in the soil. The best proof of this statement is that plants can be grown from seed to seed in synthetic chemical solutions, entirely without organic addenda of any kind, and without even the physical support furnished by soil. Such plants are as capable of supporting the growth of the animals that eat them as are "organically grown plants."

These are facts, as certain (and as tentative) as any I know in science. Yet, I think it is fair to say that they were not really believed by some students in the class who are emotionally committed to "organic gardening." This is true partially because they are in revolt against "the establishment" and the synthetic, plastic world it has created. This leads them to propose impossible additional conditions before they will believe the facts cited above. "How do you know that there aren't undiscovered vitamins or growth factors that you can't analyze for?" Of course, scientists don't know, but point to the fact that synthetic diets made of known growth factors will support normal growth of test organisms, including man, and will support reproduction of test animals, such as mice and rats, over several generations. This puts the burden of proof for other growth factors squarely in the skeptics' court, but many of them feel no necessity to prove anything; they are content to reiterate the possibility that there is more to know about diet than we now know, without being more specific and without paying regard to the usual rules of evidence.

The Organic Gardener and Anti-intellectualism

My advice to devotees of organic food is to pay the extra cost of such foods if they wish to avoid some pesticide residues and preservatives, but not to do so under the illusion that plants grown on organic media are nutritionally any better than plants given an optimum mixture of chemical fertilizers. It is doubtful whether such advice will be taken.

In a follow-up discussion, I asked how many students regularly consume vitamin pills. About one-third of the class raised their hands. I then pointed out that the body needs only minute quantities of vitamins, that a typical Yale College diet more than adequately provides the quantities needed, that excess vitamins are merely excreted, and that massive quantities of certain vitamins could actually produce harmful effects. The students listened respectfully, but later queries revealed that practically none had changed their habits and continued, often at great personal expense, to pop useless and perhaps harmful pills as part of their daily routine. They did not really counter the evidence I advanced, nor did they change their way of thinking. They continued a daily practice that was demonstrably illogical, yet somehow comfortable for them. I conclude that the Yale undergraduate, like others in the community, does not behave like a completely logical animal, cherishes his prejudices dearly, and treats science and scientists with interested attention, but not with complete trust.

25

Coda

SCIENTISTS are often criticized for doing work that in any way contributes to military weaponry or that in any way becomes subverted to antisocial uses. Why, lay people want to know, don't scientists simply abstain from doing anything that could possibly be misused and turned into counterproductive programs? The answer is, of course, that scientific discoveries are by themselves neither necessarily good nor bad for society. Einstein's discovery of the interconvertibility of matter and energy was ethically neutral. Its use by society for the generation of power through nuclear reactors could be socially beneficial, while its use to detonate an atomic bomb is obviously socially destructive. DDT is beneficial in that it kills mosquitoes that cause malaria; it is harmful in that it also destroys useful insects, like the bees that pollinate fruit orchards, and harms bird life by interfering with calcium deposition in eggshells. The list could be multiplied to include many of our most cherished modern innovations. In short, to abstain from working on subjects that might be used in any destructive way would mean to cease all scientific effort. Not only is this impractical, but it would probably be absolutely catastrophic for the future of mankind. With our increasing population and decreasing resources, we must live increasingly by our wits; in this effort the application of the newer scientific and technological information is absolutely essential for survival.

Perhaps a personal account would help to illuminate the ethical and moral quandary in which many scientists find themselves. During my graduate work on the interface between botany and chemistry, I became interested in the possible chemical regulation of the reproductive habit of plants. As I described in a previous chapter, many plants are induced to flower by the appropriate length of day, and the evidence at hand strongly suggests that the appropriate day length causes the synthesis in the leaves of a reproductive hormone, which then migrates to the bud. I noticed that in the soybean plant (which is economically important in the state of Illinois where I did the work) the onset of reproductive activity was accompanied by a decrease in the plant's vegetative growth. I reasoned that if I could find a substance that would slow down vegetative growth, I might promote the tendency to form more flowers, which would lead to more soybean pods and a bigger harvest. I decided to try to find a substance that would function as an antagonist of the plant's growth hormone, and would thus inhibit the tendency to vegetative growth. I was successful in this quest, and in my Ph.D. thesis reported that the chemical substance known as 2,3,5-triiodobenzoic acid (TIBA), when applied to soybeans growing under appropriate conditions, could greatly increase the number of flowers formed. Many years later, and unbeknownst to me, this basic finding was patented by another scientist and assigned to the International Minerals and Chemical Company of Illinois. Substantial quantities of TIBA were sold to increase soybean yields in the more northerly reaches of this crop's agricultural range until TIBA was replaced because new genetic lines worked well without it.

Anyone hearing this part of the story would conclude that my discovery of the flower-promoting action of TIBA was a socially constructive act. Yet there were subtle and unforeseen corollaries to my discovery that, many years later, caused me some embarrassing moments. For if TIBA is used at somewhat higher concentrations than I had used, it produces side effects, including the shedding of buds and even of leaves from the plants. This process of abscission can be very important in the control of plants for both productive and destructive objectives. In cotton fields, for example, where mechanical harvesters are now used in preference to hand labor, it is constructive and desirable to spray a defoliant over the plants and free the plant of unwanted leaves before the mechanical harvester goes through the field to collect the bolls. But, as Americans came to know, massive quantities of chemical defoliants were also used over the forests and farms of Vietnam for destructive purposes. I have some reason to believe that my early investigations on TIBA helped several researchers at the chemical warfare laboratories at Fort Detrick in their design and understanding of the defoliating action of chemicals of the plant hormone type. Thus, confronted by the spectacle of the misuse of my discovery, I found myself, several years ago, forced into a protest against this misuse and into a growing involvement in the social relevance of my own scientific effort.

In the decade starting with 1961, American military forces sprayed more than 100 million pounds of assorted chemicals over more than 5 million acres of South Vietnam. Substantial percentages of the hardwood forests of that country were severely damaged, with great potential economic loss to the lumbering industry of South Vietnam, one of the major sources of potential export trade. Certain important ecological niches, such as the mangroves lining the estuaries near Saigon, were permanently killed, perhaps never to recover. Crops feeding more than half a million people per year were destroyed, causing a great upset in the lives of many innocent noncombatants, including the Montagnards. When I first became aware of the extent of this program, I wrote a letter to President Johnson asking him whether he understood the broad ecological implications of the use of chemical weaponry. He replied, through an assistant secretary of state, that these weapons were being used on the advice of qualified scientists, that they produced no effects deleterious to humans, animals, or to the ecosystem, and that in any event they were the same chemicals that were being used at home in agriculture, so that their use could not be criticized. Aware that there was no really adequate scientific basis for these conclusions, I tried to involve several botanical societies in raising questions to the government, but at that time my efforts were completely unsuccessful. Later, thanks to several determined and dedicated individuals, pressure was put on many scientific organizations. Some raised their voices, the American Association for the Advancement of Science (AAAS) sponsored an investigation, and gradually the entire nature of our defoliation operations in Vietnam was revealed to the public. So ill-advised did our chemical warfare in Vietnam seem that President Nixon announced that all such operations would be phased out when existing stocks of the chemicals were gone. He also announced that he would submit to the Congress a proposal that we ratify the Geneva Protocol of 1925 forbidding the use in warfare of "asphyxiating, poisonous or other gases, and all analogous liquids, materials or devices." The additional revelation that one of the major herbicides employed in Vietnam, 2,4,5-trichlorophenoxyacetic acid (2,4,5-T), either by itself or because of a dioxin impurity it contains, can produce malformation of developing embryos in laboratory animals and possibly in humans, has led to curtailment of this product in the United States and a frantic crash program to discover whether, in fact, the chemical can ever again be used safely.

So, even a botanist, one of the last of the scientific innocents, can be involved in problems permeating every aspect of society. Many scientists are now coming to feel that they cannot surrender control of their findings to businessmen, politicians, or others for indiscriminate and unregulated use in social or military contexts. More and more, the scientist needs to concern himself with the social consequences of his discoveries. To do this he is reshaping his organizations, forming new groups, and making frequent journeys to Washington to convince legislators and government officials of the necessity of altering unwise policies. Social relevance is here to stay, and science will never be the same again.

BIBLIOGRAPHY

Introduction

Galston, A. W.; Davies, P. J.; and Satter, R. L. 1980. *The life of the green plant*, chap. 1., 3rd ed. Englewood Cliffs, N.J.: Prentice-Hall, Inc. (This book can serve as a general reference for all subjects discussed in *Green Wisdom*).

Chapter 1: New Ways to Increase Man's Food

Food and agriculture. *Scientific American*, September 1976. (An entire issue containing twelve articles devoted to an analysis of agricultural problems.)

Abelson, P. H., ed. 1970. *Food: politics, economics, nutrition and research*. Washington, D.C.: American Association for the Advancement of Science.

Chapter 2: The Water Fern-Rice Connection

Brill, W. J. Biological nitrogen fixation. *Scientific American*, March 1977.

Chapter 3: The Prodigal Leaf

Meidner, H., and Sheriff, D. W. 1976. *Water and plants*. New York: John Wiley and Sons.

Kramer, P. J. 1969. *Plant and soil water relationships*. 2nd ed. New York: McGraw-Hill and Co.

Chapter 4: The Membrane Barrier

Singer, S. J., and Nicolson, G. L. 1972. The fluid mosaic model of cell membranes. *Science* 175: 720–731.

Holter, H. How things get into cells. *Scientific American* September 1961.

Chapter 5: The Blind Staggers

Trelease, S. F., and Beath, O. A. 1949. *Selenium*. New York: Published by the Authors. (Printed at The Champlain Printers, Burlington, Vt.).

Bibliography

Chapter 6: Botanist Charles Darwin

Bates, M., and Humphrey, P. S. eds. 1956. *The Darwin reader*. New York: Chas. Scribner's Sons.

Chapter 7: Which End Is Up?

Wareing, P. F., and Phillips, I. D. J. 1978. *The control of growth and differentiation in plants*. New York: Pergamon Press.

Chapter 8: Rotten Apples and Ripe Bananas

Abeles, F. B. 1973. *Ethylene in plant biology*. New York: Academic Press.
Biale, J. B. The ripening of fruit. *Scientific American*, May 1954.

Chapter 9: Turning Plants Off and On

Moore, T. C. 1979. *Biochemistry and physiology of plant hormones*. New York: Springer Verlag.

Chapter 10: Sex and the Soybean

Smith, H. 1975. *Phytochrome and photomorphogenesis*. New York: McGraw Hill and Co.
Hillman, W. S. 1962. *The physiology of flowering*. New York: Holt, Rinehart and Winston.
Salisbury, F. B. The flowering process. *Scientific American* April 1958.

Chapter 11: Plants Have a Few Tricks Too

Deverall, B. J. 1977. *Defense mechanisms of plants*. New York: Cambridge University Press.
Harborne, J. B. 1977. *Introduction to ecological biochemistry*. New York: Academic Press.

Chapter 12: The Language of the Leaves

Haupt, W., and Feinleib, M. E. eds. 1979. *Plant movements*. Encyclopaedia of plant physiology New Series, vol. 7. Berlin–Heidelberg–New York: Springer Verlag.
Bünning, E. 1979. *The physiological clock*. Berlin–Heidelberg–New York: Springer Verlag.

Bibliography

Chapter 13: A Basic Unity of Life

Smith, H., ed. 1976. *Light and plant development*. London–Boston: Butterworth's.

Chapter 14: The Limits of Plant Power

Lund, E. J. 1947. *Bioelectric fields and growth*. Austin: University of Texas Press.
Scott, B. I. H. Electricity in plants. *Scientific American*, October 1962.

Chapter 15: The Immortal Carrot

Street, H. E., ed. 1973. *Plant cell and tissue culture*. Berkeley and Los Angeles: University of California Press.

Chapter 16: The Naked Cell

Hughes, K. W., et al. eds. 1978. Propagation of higher plants through tissue culture. Technical Information Center, U. S. Department of Energy.

Chapter 17: Molding New Plants

Street, H. E., ed. 1974. *Tissue culture and plant science*. New York: Academic Press.

Chapter 18: Here Come the Clones

References for chapters 15–17.

Chapter 19: Plant Cancer

Braun, A. C., and Stonier, T. 1958. Morphology and physiology of plant tumors. *Protoplasmatologia* 10: 1–93.

Chapter 20: In Search of the Antiaging Cocktail

Hayflick, L. Human cells and aging. *Scientific American*, March 1968.

Chapter 21: A Living Fossil

Siegel, S. M., and Siegel, B. Z. 1968. A living organism morphologically comparable to the pre-cambrian genus *Kakabekia*. *American Journal of Botany* 55: 684–687.

Bibliography

Chapter 22: Guayule Bounces Back

Vietmeyer, N. 1977. *Guayule, an alternative source of rubber*. Washington, D.C.: National Academy of Sciences.

Chapter 23: How Safe Should Safe Be?

Whiteside, T. 1978. *The pendulum and the toxic cloud*. New Haven: Yale University Press.

Ehrlich, P. R.; Ehrlich, A. H.; and Holdren, J. P. 1977. *Ecoscience: population, resources, environment*. San Francisco: W. H. Freeman and Co.

Chapter 24: Organic Gardening and Anti-intellectualism

Wortman, S., et al. Food and agriculture. *Scientific American*, September 1976.

Chapter 25: Coda

Galston, A. W. 1972. Science and social responsibility: A case history. *Annals of the New York Academy of Sciences* 196: 223–235.

INDEX

abnormal growths, 149–50, 154
abscisic acid (ABA), 28, 29, 78–79
abscisin, *see* abscisic acid
abscission, *see* defoliation
abscission layer, 71
absorption, 19, 25, 36
ACC in ethylene synthesis, 72
acetone, 172
acetylcholine, 107–8
acetylene, 68–69
acidity in *Azolla* plants, 22–23
active transport, 111
adaptation, 29–30, 42
Addicott, Frederick, 77
adenosine triphosphate (ATP), 36, 38, 72, 101
Agate soybean, 82
aging, *see* life span
agriculture: and disease-resistant crops, 93; and ethylene derivatives, 70, 71; and genetics, 118, 140; and growth-regulatory chemicals, 66; and hormone use, 56, 58, 71; and propagation from single cell, 124; and synthetic chemicals, 181–82; and technology, 117
Agrobacterium tumefaciens, 149
Albizzia (silk tree), 100, 101, 102, 104–8
alga, 19, 169
algal plant cells, 33
alkali disease, 39, 41, 42, 43
alkalinity, 167–68
alkaloids, 52–53, 92
alpha hydroxy pyridine methane sulfonic acid (HPMS), 16
altitude of *Kakabekia* growth, 168
American Association for the Advancement of Science (AAAS), 197
amino acid: and ethylene production, 72; and mutant plants, 124–

25; and nitrogen fixation, 80–81; and selenium, 40; and sulfur, 43
ammonia, 167
ammonium carbonate, 53
amoeba, 128
amylopectin, 183
amylose, 183
Anabaena (blue-green alga), 19–22
Anabaena-Azolla-rice mutualism, 18–22
animals: cell culture in, 127–28; electrophysiology in, 109–10; and nuclear fusion, 128; organic farming, 190, 191, 192; and pesticides, 190; plant defense against, 90, 91
anthers, 124, 135
antiaging cocktail, 159, 161
antioxidant and radiation, 159
apical hook, and ethylene synthesis, 72
apple, 67–70
aqueous pathways, 33–34, 35
Arabidopsis, 137
arid climate, 29–30, 175
artificial ripening, and ethylene, 70
Astragalus, 39–46
asymmetric growth, *see* geotropism; phototropism; tropistic responses
atomic bomb, and antisocial science, 194
ATP, *see* adenosine triphosphate
atrazine, 182–86
autotroph, 190
auxin: basipetal movement of, 56; and crown gall, 151; and electrical potential, 113; and haploid plants, 124; and plant growth, 113, 119–20; and plant movement, 56; and trans-cinnamic acid, 178; *see also* auxin-cytokinin ratio

Index

auxin-cytokinin ratio, 120, 124, 127
Avogadro's number, $7n$
Azolla pinnata, *see* water fern
Aztecs, and rubber production, 171

bacteria, 80, 89, 149–51, 168; and DNA, 137; and genes, 143, 150; and longevity, 155; and transduction, 136
bacterial cell chromosome, 144
bagasse, 172, 176
balanced growth, and hormones, 148–49
bananas, and effect of ethylene, 70
Barghoorn, Elso, 166
Batorawka soybean, 82
bees, and danger of DDT, 194
Beagle, 51
Berzelius, Jöns Jacob, 41
Biloxi soybean, 82
biochemical mechanisms, 104–8, 147
bioelectric properties, and light, 113, 114
birch tree winter bud formation, 75, 77
birds, and danger from DDT, 194
bitter chemicals for defense, 92
blade of leaf, and position, 97
blastomeres, in cloning, 141
blastula, in cloning, 141
blind staggers, 39, 41–42, 43–45
blood serum, 127–28
blue-green algae, in rice paddies, 19, 21
blue pigment, *see* phytochrome
bone calcification, and collagen, 161
Bonner, James, 178
botany: and ethics, 197; and social consequence of discovery, 197, 198
Boysen-Jensen, Peter, 56
breeding: and disease resistance, 118; and guayule improvement, 176; in haploid plants, 124–25; and photorespiration, 17; and protoplast manipulation, 127
breeding cycle, in plants, 119

bristlecone pines, and longevity, 156
bromine water, and hydrocarbons, 69
brown seaweed, 113–14
buds: and auxin-cytokinin ratio, 120, 127; and sucrose, 87, 88; and TIBA influence on shedding, 196
bull-horn (desert *Acacia*), 91
Bünning, Erwin, 98, 105
Burr, Harold S., 112
Butler, Samuel, 49
butternut, winter buds on, 76
butylene, 68

cabbage, and clubroot malformation, 149
calcium, 114, 129
callus, 120, 127
calories of energy, in nuclear furnace, 7
cambium, assymetrical growth of, 65
cancer: and crown gall, 149; and herbicides, 182
cancer research, 37
carbohydrates, in photosynthesis, 80
carbolic acid, and phenol, 92
carbon, 8, 32, 119
carbon dioxide, 13–15 passim, 24, 25; and *Anabaena* alga, 19; and climacteric in fruit, 70; as ethylene antagonist, 69; fixation of, 14, 15, 16, 17; and open pore space, 28; in photosynthesis 3, 4, 13 19, 24–25, 80; and stomatal opening, 29
cardboard, by-product of guayule, 176
Carrel, Alexis, 159
carrot: life cycle of, 119, 121, 122–23; tissue culture of, 119, 159
cell: and cloning, 143; and cultures, 128, 160; and differentiation, 129; and DNA receptor, 142; fusion in, 128, 131; mutation and aging in, 161; radiation of, 159

Index

cell layers, 32
cell membrane, 31–38, 105, 111, 128, 130; *see also* membrane; surface membrane
cell transformation, 130
cellular differentiation, 141, 147, 148
cellular fusion, 128
cellular salt changes, 111
cellulase, and mold secretions, 126
cellulose, 14
cellulose cell wall, 126, 127
cell wall, and thickness of guard cells, 26–27
centrifugation, and chromosomes, 142–43
centrifugal force, and plant direction, 60, 61, 62
centrifuged membrane "pellet," 32
chemicals: as defense mechanisms, 92–93; and regulation of reproduction, 195; and rubber technology improvement, 176
chemical analysis, of cell membrane, 32
chemical mutagens, 184
chemical pesticides, and pathogens, 118
Chihuahua Desert, 171
chloride, and human nerve cell, 110
chloride ion, and plant movement, 101
chloroethanephosphonic acid, 71
chlorophyll, 8, 13, 24, 80
chloroplast, 8, 32, 36, 80, 117
chromosomal errors, 161
chromosomal insertion, 143
chromosomes: and Feulgen stain, 169; and colchicine, 135–36; division of, 125; and genetic transformation detection, 124; and isolation by centrifugation, 142–43
chrysanthemum defoliation, 68, 71
circadian rhythm: Bünnings's studies on, 105; and electric potential, 111–12; and leaf closure, 102–3, 107; and photoperiod, 83; and phytochrome, 75; in plant movement, 101

circumnutation, 53, 55
climacteric in fruit, 70
cloning, 141–46
Clostridium, 168
clubroot, 149
colchicine, 125, 126
Colchicum autumnale, and colchicine, 125
collagen, and aging, 161
compression wood, 65
cones, in eye membranes, 37
conifer, and adjustment of growth direction, 65
connective tissue, 161
contact inhibition, 37
Continental-Mexican Rubber Company, 174
corn, 24, 25, 28, 182–83
corn blight, 118
cosmic ray, and mutations, 184
cotton boll, shedding studies on, 77
cowslip, and dioecious tendencies, 52
crop yield, 16–17; and cereal grains, 117; in chemical versus organic farming, 190–191; and measurement of plant productivity, 14; and nitrogen's effect on, 131; pathogen susceptibility with increase in, 118; and photorespiration, 15; rubber formation, 176; soybean and, 81; time needed for development of increase in, 141; variations in farm crops of, 13; in water fern-rice paddies, 19
cross-fertilization, 51
crossing over, 144–45
crown gall, 149–54
cryophilic plant strain, 168
culture: of cancer cells, 159–60; of carrot tissue, 119; of cells, 120, 128; of protoplast, 127; of sycamore, 137
curvature: *see* growth curvature
cuticle, 26
cuticle wax, on guayule leaves, 176
cyclic adenylic acid, 108
cysteine (cystine), and sulfur, 43
cytokinin, 119–20, 124, 127, 151

Index

dairy industry, and cloning, 142
Darwin, Charles, 49–58; *Movements and Habits of Climbing Plants*, 53; *Origin of Species*, 51; *Power of Movement in Plants*, 53; *Various Contrivances by which Orchids are Fertilized by Insects*, 51
Darwin, Erasmus, 49
Darwin, Francis, 53
daylight, *see* length of day
DDT and controversy, 182, 194
death statistics, 156
defense mechanisms, 89, 90–93, 118
defoliation: in chrysanthemum, 68, 71; and flowering, 88; and hormones, 58; and TIBA, 196; and unsaturated hydrocarbons, 68–69
degeneration in humans, 156
dehydration, 78
deoxyribonucleic acid (DNA): aging and, 161; and atrazine, 186; and base-pair substitution, 185–86; and cell as receptor, 142; and cell transformation, 130, 136; and crossing over, 144–45; Feulgen stain, 169; and gene switching, 148; and genetic experiments with, 136–39, 142, 144–45; and genotype formation, 130; and herbicidal mutagens, 184, 185–86; and ingestion, 37; and microbial cells, 130; and recent research, 31; transfer from virus to cell, 137; in *Rhizobium*, 81
depolarization, 111
desiccation, prevention of, 29, 78
detoxification, in humans, 186
diabetes, genetically engineered cure, 145
diatoms, 169
diet, and life-span, 158
differential permeability, 34
differentiation, 114, 128, 129
diffusion, 34–35, 38
dimorphic flowers, 51–52
dioecious tendencies, 52
dioxin, and embryo malformation, 197
diploids, 124, 125, 135–36

directionality, 35, 59–66
disease resistance, 118; *see also* defense mechanisms
DNA, *see* deoxyribonucleic acid
dogs, and genetic manipulation, 140
donor leaf, in grafting, 86
dormancy, 58, 74–75, 77, 78–79
dormin, 77
down-up axis, 61
Doy, Colin, 137
Drosera (sundew), 52, 90, as insectivorous plants, 52
dual solubility, in cell membrane, 32–33

ecology, 190, 197
eggshells, and DDT, 194
Eisenhower, Dwight D., 175
Einstein equation, thermonuclear fusion, 6–7
Einstein's theories, 194
electrical bonds, and energy, 6
electrical changes, in animals, 110
electrical fields and gradients, 112–13
electrical impulse and its effect on plants, 38, 107, 110
electrode implants and motor cells, 111–12
electrophysiology of animals, 109
electrophysiology of plants, 109–114; *see also* transverse electrical potential
Emergency Rubber Project, 173
endogenous rhythm, *see* circadian rhythm
energy, 6–9 passim, 36; and cell membrane permeability, 35; and conversion in chloroplasts and mitochondria, 36; enzymes and, 169; in plant movement, 101; and photorespiration, 15; and photosynthesis, 3, 24; and respiration, 14; and sources in the plant, 6, 14
environmentalists, 182
environmental manipulation, 16

Index

Environmental Protection Agency, 183, 186
enzymes: as catalysts, 147; and defense mechanisms, 93; and digestion of cell walls, 126; and energy, 169; and genetic messengers, 147; and insectivorous plants, 52; and nitrogen fixation, 81; and plant synthesis of ethylene, 72
epidermis, and leaf structure, 26
epinephrine in plants, 107–8
epiphytes, 90
Ethephon, 71
ethylene, 67–73
evaporation, 24, 29–30
evaporative cooling, and transpiration of leaves, 25
evolution, xi, 51
evolutionary theory, of Charles Darwin, 51

and ethylene release in pineapples, 71; and hormones, 58; and length of day, 195; light requirements for, 86–87; photoperiod control of, 82; for reproduction, 80; and reproductive hormone synthesis, 195; in soybean, 82
Food and Drug Administration, 186
food preservatives, 190
food production, see crop yield
forests, 26
fossil plant, see Kakabekia
free radicals, and cell radiation, 159
fruit flies, 68, 158
fruit ripening, 58, 67, 69–71
fruit-set, and photoperiod, 58, 82
Fucus (brown seaweed), 113–14
fuels, and energy, 6
fungal toxicity, 92, 93
fungus, 67, 89, 92, 93
fusion, 5, 128–29

far-red light, 106–7, 112
fat, 14, 32, 33–34, 176
ferns, 18; and absence of seed production, 23; see also water fern
Ferocactus, 30
fertilization, 131; and conventional genetics, 131
fertilized egg as DNA receptor, 142
fertilizer, 23, 81
fertilizer salts, and pollution, 133
Feulgen stain, 169
fibroblast: and cell culture, 128; and "hourglass" theory of aging, 160
filamentous fungus, 93
Fitz-Roy, Captain, 51
flaccidity, in plant movement, 100
fleshy stem, 30
florigen: and flowering plants, 86; and isolation attempts, 88; in leaves, 83; and movement through plant, 83; production of, 98; in soybean, 87
flowering: in the carrot, 122–23; and circadian rhythms, 103, 112;

galactose, 144
galactosemia, 143–44
galactose-1-phosphate, 144, 145
gall, on oak trees, 149
Galston, Arthur, and identification of trans-cinnamic acid, 178; and soybean experiments, 195–98
Galton, Sir Francis, 49
Galvani, Luigi, 110
Gautheret, 119, 127
gender, and life span, 158
genes, 143, 147
gene switching, 148, 150
genetic background, and life span, 157
genetic change, 136, 139; induction of, 125; potential benefits of, 139; and transduction in bacteria, 136
genetic coding, 144
genetic deficiencies, in Arabidopsis, 137
genetic engineering, 140, 141–46, 176

Index

geneticists, and disease-resistance, 93, 118

genetic messengers, and enzymes, 147

genetic mistakes, and aging, 161

genetics, conventional practice of, 131; and new strain development, 118

genetic selection, and crop yield, 17

genetic switching, mechanism for, 148

genetic transfer, between cells, 37

Geneva Protocol of 1925, 197

genomes, 128–29

Gentile, James M., 182, 183–84, 185–86

geotropism, 62; Darwin's discovery of, 55, 56; mechanism of, 63, 64; and root growth direction, 60; and statolith theory, 61; and surface membrane, 63; in Swedish Ivy, 57

germicidal compounds, 92

germination, and abscisic acid, 79

gibberellins, 87, 88

glucose, from lactose, 144

glycerol, structure of, 32

glycolic acid oxidation, 15–16

grafting: and crown gall, 152–53; and florigen, 83, 86

Grahm, Lennart, 113

grasses, and tropical crop yields, 13

gravity: and geotropic responses, 64; and growth curvature, 112–13; and movement of plants, 53, 55

Gray, Asa, 53, 54

green plant: as autotroph, 190; for energy and food, 6, 9; as a fuel source, 6; and selenite salt metabolism, 43; as solar energy converter, 8

growth curvature: and electrophysiology, 110, 112–13; see also geotropism; phototropism

growth hormone: in cancer-like cells, 150; chemical interference with, 66

growth inhibitors, 63, 64, 75, 77

growth patterns, 112, 113, 119–20

growth polarization, 114

growth-promoting hormone, 63

growth rate, 110, 168

growth-regulating hormone: and transverse electrical potential, 63

guard cells, 26–27, 28, 78–79

guayule, 171–76, 177; bagasse and by products of, 172, 176; genetic engineering of, 176; and rubber plant, 171; rubber production from, 175; as source of rubber, 174; and World War II rubber production, 173; see also rubber; rubber production

haploids, 124–25, 136–37, 184

hardwood tree, and adjustment in direction of growth, 66

hardwood forests, and defoliant, 196–97

hawthorn, and defense mechanism, 91

Hayflick, Leonard, 160

heart, and muscle aging, 161

heat buildup, in leaves, 25

HeLa cancer strain, 160

helium, 5–7

heme enzyme, and energy production, 169

Henslow, J. S., 51

herbicide: and cancer, 182; and embryo malformation, 198; safety testing of, 182, 187; and toxicity, 183

herbicide mutagenicity, 182, 183–86

hereditary material, DNA as, 137

hermaphrodites, 52

Hertz, C. H., 113

Hess, Dieter, 137

Hevea (rubber tree), 173, 175–76, 177–78

Hildebrandt, Albert C., 134–35; obtaining virus free plant strains, 134–35

hormonally controlled correlations, 49

hormone, auxin, 56, 119–20; and balanced development, 149; and

Index

cell permeability, 35; cyclic adenylic acid, 108; and cytokinin, 119–20; definition of, 73; effect on potassium chloride level, 28; ethylene described as, 72–73; and growth antagonist, 195–96; indoleacetic acid, 108; and mobile inducer hormone, 148–49; and plant tissue growth, 119; and starch grains, 64; studies of, in botany, 56; and target tissue, 148; see also auxin
"hourglass" theory of aging, 160
HPMS, see alpha hydroxy pyridine methane sulfonic acid, 16
humanistic biology courses, 188–89
hybrid, 52, 128, 138
hybrid corn, 140
hybridization, in mustard, 129
hydrocarbons, 67, 68–69
hydrogen, 5–7
hydrogen ion pumps, 113
hydrophilic regions, of protein, 33
hydrophobic regions, of protein, 33

iron selenide, 41
irrigation, 29
isoprene, molecular structure of, 176

Jaffe, Lionel, 113
Jaffe, Mark, 107, 108
Johnson, C. B., 137
Johnson, Lyndon B., on use of chemical warfare, 197
Jupiter, and ammonia in atmosphere, 167

Kakabekia, 166–70; discovery of, 166; and Feulgen stain, 169; growth rate of, 168; genetic variants of, 161; is indigenous to, 169–70; and silica cell walls, 169; and sodium metasilicate, 168
Kakabekia barghoorniana, 161
Kakabekia umbellata, 161

immature embryos and cultures, 127–28
immunology, and cancer research, 37
inactive virus, and cell cultures, 128
Indian reservations, and guayule production, 174
indoleacetic acid, 108
ingestion, 37, 130; and surface membrane, 37
insectivorous plants, 52–53
insects, and Azolla, 22, 23; plant defense against, 90, 92; and plant malformation, 149
internal clock, see circadian rhythm
International Minerals and Chemical Company of Illinois, and use of TIBA, 195
iodine-potassium iodide reagent, in test for herbicidal mutagenicity, 183, 184
ion movement, and plant direction, 64

Lacks, Henrietta, 160
lactose, and galactosemia, 144
lamina in daytime position, 93
Lansing, Albert, 157
leaf: in arid climate 29; circadian rhythm in, 105; closure and phytochrome, 106–8; dorsal surface of, 111; electrical signals and closure of, 111, 112; energy loss in, 25; and florigen synthesis, 86; folding of, 97–99, 101; function of, 24; and growth curvature, 112–13; guard cell of, 26–27; and oxygen loss, 25; photoperiod perception, 82; and phytochrome-circadian rhythm interplay, 75; and reproductive hormone synthesis, 195; salt concentration, and effect on, 101; stomata in, 26–30; structure of, 26–27; and transpiration, 26; see also leaf movement; leaves, photosynthesis
leaf movement: and circadian

Index

leaf movement *(continued)*
rhythm, 101; in leguminous plants, 99–100; light controlled, 100–101, 102, 103, 108; mechanism of, 99–103, 104–8
leaky membranes, and crown gall, 151
leaves: air movement and, 25; awake position of, 100; and defense mechanisms, 90; function of, 149; sleep movements of, 97–103; sugar production in, 4; of the water fern, and *Anabaena*, 19; *see also* leaf; leaf movement; photosynthesis
Ledoux, Lucien, 137
leguminous plants, 41, 99–101
length of day: and reproductive hormone release; and timing mechanisms, 74, 75
life cycle: of carrot, 119; of guayule rubber shrub, 176; of pathogenic mutations, 141
life span, 156–58; of cells removed from body, 159; changes in, 159; and collagen, 161; influences on, 159; and mutation in cells, 161; studies of, 160
light: absorption of, 25, 98–99, 106, 108; and bipolarity, 114; and carbon dioxide, 15, 29; Einstein's equation and velocity of, 7; exposure time of, and winter preparation, 75; and flowering requirements, 86–87; and growth curvatures, 112–13; and leaflet movement, 105; and movement of plants, 53, 55, 100–1, 102, 103; and photoinductive daylight, 86; and photosynthesis, 3, 13; and polarized eggs, 114; and potassium chloride level, 28; protection of *Azolla* and intensity of, 23; and stomata, 28, 29; and transmission of stimulus, 55, 56
light-dark transitions, 111–12
lignin, 65
Lindbergh, Charles, 159
Linnean Society, 53

Linum flavum (flax), and dimorphism, 52
lipids, 32
livestock, and selenium related disease, 39, 41–45
locoweed, 44
long-day plants, and leaf folding, 99
longevity, *see* life span
Lund, E. J., 112
lungs, and muscle aging, 161
Lythrum (loosestrife), 52

magnolia, and winter buds, 76
malaria, and DDT, 194
malformations, 149, 198
malignant tumors, 149–50, 151
mannitol, 127
mass, and Einstein's equation, 7
measurement of cell layers, 32
meiosis, 135
membrane: as barrier, 33; and changes in animal nerve cells, 110–11; and permeability related to leaf movement, 103 (*see also* permeability of the cell membrane); specialized, 37–38; structure of, 8, 36; variable quality of, 36
metabolism, and life span, 159
methionine 43, 72
Mexico, and guayule rubber, 175
microbes, defense against, 90, 92–93
microbial cells, DNA introduced into, 130
microbial test organisms, 185
microorganisms, in soil, 191
microspore, and propagation, 124
Mimosa, 90, 111
minerals, 19, 26, 81, 119
mitochondria, 32, 36
mobile inducer hormone, 148
molds, 92, 126
molecular genetics, 141
mosaic membrane quality, 33–34
Monilinia, non-pathogenic fungus, 93
mosquito fern, *see* water fern

Index

mother's age, and life span, 157–58
motor cells, 103, 104, 111–12
movement, *see* plant movement; sleep movement
Movements and Habits of Climbing Plants, The (Charles Darwin), 53
multicellular organisms, 147
muscle cells, and aging, 161
mustard, 92, 129, 137
mutagen: and atrazine, 183, 184, 185; and genetic change in DNA, 185–86; and haploid plants, 124–25, 139; and new strain development, 118
mutating pathogens, *see* pathogens
mutation: amino acids and; 124–25; in diploids, 124; effect on plant, 136; from herbicide use, 182; and stain test for waxy corn, 184

National Academy of Sciences, 174, 175
National Institute of Environmental Health Sciences, 183
National Institutes of Health, 183, 186
National Aeronautics and Space Administration, 166; and studies of organisms under stress, 166
natural gas, 68–69
natural rubber, 177
neural impulse, and acetylcholine, 107
neuron behavior, movement mechanism, 112
neurotransmitting substances, 107–8
nerve cells, 110–11, 160
Newman, Ian, 112
new strain, 118, 124, 136
nitrogen, 19–22, 131
nitrogen fixation: and *Anabaena-Azolla* combination, 22; by nodules of legumes, 81, 131–32; process of, 80–81; and protoplast fusion, 133; and *Rhizobium*, 80;

for rice plants, 21; by symbiotic nitrogen fixers, 131–32
nitrogenous fertilizer and soybeans, 81
Nitsch, Colette, 135
Nixon, Richard M., on chemical warfare, 197
Nobécourt, 119, 127
nodules, and nitrogen fixation, 81, 131
norepinephrine, 108
North Vietnam, rice production in, 19
Norway maple, and winter buds, 76
nuclear energy controversy, 7
nuclear reactors, 194
nuclear transplantation, 142, 143
nucleic acid, and radiation, 159
nucleus, 32
nutrition, and inorganic versus organic gardening, 190–92
nutrition, and life span, 159
nyctinastic closure, *see* sleep movement

oak, gall on, 149
obligate ammonophile, and *Kakabekia*, 167
onion, 92
oils and shellac, from guayule plant, 176
open pore space, in leaf, 26, 28
organelles, 32, 36
organic acids, and mitochondria, 36
organic farming, 186, 190
organic matter: and photosynthesis, 3, 13, 14; and the soil, 191–92; storage in vacuoles, 28
organic versus inorganic gardening, 190–92
Origin of Species, The (Charles Darwin), 51
osmosis, and protoplast, 126–27
osmotic concentration, 28, 100
Overton, 33–34
oxidation, 3, 14, 36

Index

oxygen: and crop yield, 17; *Kaka-bekia* cultures and, 168; leaf loss of, 25; open pore space and, 28; and photorespiration, 17; in photosynthesis, 4, 13, 24

paper, as by-product of guayule, 176
parasexual cell fusion, 128, 129
pasteurizing to prevent crown gall, 151
pathogenic organisms, 141, 145
pathogens, susceptibility to, 118
pea pod, 93
pea seedling, and apical hook, 72
Pearl, Raymond, 157
pectins, 126
Peking soybean, flowering in, 82
Pelvetia (brown seaweed), 113–14
penicillin, and genetic engineering, 140
permeability of the cell membrane, 33–37, 103, 128, 130
peroxidase, 93
pesticides, 23, 118, 187, 190
petioles, and geotropism, 64
Petri dish, and cell cultures, 128
petunia, 129, 137
phenols, 92, 93
phenol-oxidizing enzyme, 169
phloem tissue, and florigen movement, 86
phosphates in glycerol molecule, 32
phosphate bonds, and energy release, 36
phospholipids, 32, 33
photoinductive daylight, 86
photoperiod: in carrot, 123; and circadian rhythms, 83; and flowering, 82; and plant sensitivity, 98–99; in soybean, 82–83, 84, 85, 86–87
photophile, and circadian cycle, 105
photorespiration, 14–17; and carbon dioxide, 14–15
photosynthesis, 3, 4, 13–14; and agriculture, 8; and *Anabaena* alga,

19; and carbohydrates, 80; and carbon-dioxide, 13, 19, 24–25, 80; carbon fixation in, 8; cellulose as end product of, 14; and chlorophyll, 13, 24, 80; and chloroplasts, 80; dependency of all living matter on, 117; and energy, 3, 24; and leaves, 24, 98; and photorespiration, 15; rate of, 14; and relation to respiration, 14; and soybean vegetation, 80–81; stomata in, 18–19, 29
phototropism, 55–56, 57, 60
phytoalexins, 92, 93; for plant defense, 92
phytochrome: and acetylcholine, 107; discovery of, 82; as ethylene synthesis regulator, 72; light sensitivity, 75, 82, 105; mechanism of movement, 104–8; and plant pigment, 82; potassium movement, 107; and red and far-red light, 106, 112; and flower induction, 71
pisatin, 93
pituitary dysfunction, 145
plant, industrial, 3, 6
plant-herbicide interaction, 185
plant movement, 56, 99–103, 105; *see also* sleep movement
plant pigment, 72, 82, 137
plant preparation for winter, 75
plant regeneration, 129
plasmid, 81, 144, 145, 151
Plewa, Michael J., 182–86
Plinian Society, Charles Darwin's association with, 50
polarized egg, and light, 114
pollen: and genetic change, 125; haploid derived from, 136; herbicidal mutagenicity, 183, 184; and propagation, 124; and virus free plants, 135
pollen tube, 131; and incompatibility with stigma, 131
pollination, and conventional genetics, 131
polyethylene glycol, 128, 129
polymerized isoprene, 176
polypoidy, and colchicine, 125

Index

population profile, 156
potassium, 101, 104–5, 107, 110–11
potassium chloride, 28
potassium iodide, as stain, 183
potatoes, 124, 129, 131
Power of Movement in Plants, The (Charles Darwin), 53
prehistoric period, and atmospheric ammonia, 167
preventive medicine and life span, 156
primrose, dioecious tendencies in, 52
Primula, dimorphism in, 51–52
prokaryote, and DNA distribution, 169
propagation from single cells, 120
propylene, 68
protective mechanisms, *see* defense mechanisms
protein, in cell layers, 32; and cell membrane permeability, 34–35, 36, 128; and contact inhibition, 37; and electrophysiology, 111; gene specificity of, 148; genetic coding for, 144; and genetic messengers, 147; from guayule seeds, 176; and insectivorous plants, 52; methionine in, 72; and nitrogen fixation, 80–81; in the organelle membrane, 36; and photosynthesis, 14; sulfur in, 43; and virus, 128
protein synthesis, recent research, 31
protoplast, 126–27; and fusion, 129, 133, 138; and hybrid production, 138, 139; and hypothetical pomato, 132; and ingestion, 130
public health, 187
pulvinus: and acetylcholine, 107; in *Albizzia*, 100; and movement, 99, 101, 104–8

Racusen, Richard, 111–12
radiant energy absorption, 25
radiation: haploid plants, 124–25; and life span, 158–59; and new

strain development, 118; and receptor cell, 142; and solar energy, 8
radiative heat loss, and leaf folding, 98
radicle, and tropisms, 55–56
radioactive experimentation, 72
rate of growth, and tropisms, 60
rate of photosynthesis, 14, 16, 17
reaction wood, and tropisms, 65
red blood cells, 32, 161
red light, 106–7, 112
redwood aging, 156
reflection, and leaf energy loss, 25
regeneration, 120, 127, 129, 135
regulatory physiology, 78–79
reproductive hormone migration, 195
research on cell membranes, 31
resin, 172
renewable energy source, 8
repressor molecule, 148
reproduction, 19, 80; *see also* florigen; flowering
reradiation, in the leaf, 29
resin, 172, 176
respiration, 13, 14, 24
response patterns, and evolution, *xi*
retina, 37
Rhizobium, 80, 81
rhizoid, brown seaweed, 113–14
rhythmic leaf closure, 103
ribonucleic acid (RNA), 31, 130
rice, 19, 20, 21; and blue-green algae in paddies, 19, 21; and genetic improvements, 140; in North Vietnam, 19
rice plant, need for fixed nitrogen, 21; and cultivation with water fern, 19–23
rice wilt, and mutating pathogens, 118
ripening, and ethylene, 69–71
RNA, *see* ribonucleic acid
rods, 37
root: and arid conditions, 29; and directionality, 59; and electrical impulse, 107, 110; and far-red light, 107; function of, 148; growth direction of, 59; hor-

Index

root *(continued)*
   mone effect on, 120; *Rhizobium* in, 81; and sugar, 4
   root formation, and high auxin-cytokinin ratio, 127
   root hair, and nitrogen fixation, 81
   root tips, and electrical charges, 107
rotifers, 157
rotting: and ethylene production, 67, 68; prevention of, 69–70
rubber production: from guayule, 171–72, 175, 176; from *Hevea*, 173, 175–76, 177–78; mechanism for, 178; natural rubber, 177; polymer construction of, 176, 177; from Russian dandelion, 173

Sachs, Julius, 53
S-adenosylmethionine (SAM), 72
safety, 8, 181, 184
salt: accumulation and diffusion, 38; and content of human nerves, 110; and insectivorous plants, 52; and membrane permeability, 103; and motor cells, 111; and plant movement, 99, 100–101; and tissue growth, 119; and vacuoles, 28
sap movement in transpiration, 26
Satter, Ruth, 111–12
saturated hydrocarbons, 69
Schrank, A. R., 112–13
science, 189, 194
scientists, 189, 190
scotophile, and circadian rhythm, 105
screening programs, 181, 187
secondary tumors, and bacteria, 150–51
seed: conventional genetics and harvesting of, 131; dehydration, 79; development of, and oxygen, 17; and dormancy, 77; ferns and absence of, 23; production of, in the carrot, 122–23
selection, 140, 176
seleniferous plants, 41–45
seleniferous rock, 42

selenite salt, 42
selenium, 41, 43; and affinity of toxic plants, 41; and amino acids, 40; and *Astragalus*, 41, 42, 43; diseases related to (*see* alkali disease, blind staggers); in human diet, 46; and livestock, 43, 44; plants affected by, 42; toxic effect of, 40; and *Xylorrhiza*, 40, 41
seleno-cysteine, toxicity of, 43
seleno-methionine, toxicity of, 43
senescence, 22, 77
serotonin, 108
sex organ development, 80
sexual hybridization, 140
sexuality, and flowering, 82
shedding, and the cotton boll, 77; and TIBA, 196
sheep, 39, 44
short-day treatment, and flowering, 98
shrinking, and plant movement, 99
Siegel, Barbara, 166, 169–70
Siegel, Sanford, 166, 169–70
silica, and fossil cell walls, 169
single cell, 120, 125
slash-and-burn agriculture, 26; and mineral restoration, 26
sleep movement, 97–103, 111
sodium, 107, 110–11
sodium metasilicate, and *Kakabekia*, 168
Sokal, Robert, 158
soil, 167–68, 191
Solanaceae, 124, 131
solar energy: as basis for life, 117; and fusion, 5; and nuclear energy controversy, 7; in photosynthesis, 3, 4; and thermonuclear reactions, 5, 6–7, 8
solar radiation in water fern–rice paddies, 19
somatic hybridization, 129, 131, 132
sorghum, 13
soybean: and carbon dioxide fixation, 13; chloroplast in, 80; and crop yield, 13, 15; experiments with, 195–98; florigen buildup of, 87; flowering in, 82, 84, 87; and

Index

grafting, 83, 85; nitrogen fixation in, 80–81; photoperiod in, 82; and synthetic growth regulator, 66; vegetative process in, 80–81
staghorn sumac with buds, 76
stain reactions in waxy corn, 184
starch, and photosynthesis, 14
statolith, 63, 64
statolith theory, 61; and gravitational response, 61
stem: in arid zone plant, 30; chemical polarity of hormones in, 66; direction of growth of, 59; and electric impulse, 110; sugar in, 4; virus free, 135; woody tissues in, 26
steroid hormone, 145
stigma, and pollen tube, 131
stomata, 27–30; and abscisic acid, 78–79; in photosynthesis, 18–19; transpiration through, 26; in water fern, 18; in xerophytes, 30
strangler fig, 90; defense mechanisms in, 90
streptomycin, 140
stress, studies of organisms under, 166
sucrose: and budding, 196; and cell membrane permeability, 35; and flowering, 88; and plant tissue growth, 119
sugar: green plant storage of, 24; and oxidation in mitochondria, 36; and photosynthesis, 3, 4, 13–14, 24; in respiration, 14; see also sucrose
sugarcane, crop yield of, 13
sulfur: in amino acids cysteine and methionine, 43; and selenium displacement, 40, 43
sun, as a thermonuclear reactor, 4–6; see also solar energy
surface membranes, 37, 63
Swedish ivy, and tropism, 57
swelling, and plant movement, 99
sycamore, and genetic change, 137
symbiosis, and nitrogen fixation, 81, 131
synchronous replication of DNA, 144

synthetic chemicals, 181–82, 183
synthetic growth regulator, 66
synthetic hormones and ethylene, 71
synthetic rubber: cis- and trans-configurations, 177; and petroleum industry, 173–74

tactile sensitivity, 54, 90
tannin, as defense mechanism, 92
target tissue, and hormones, 148
temperatures: and crop yield, 16; and crown gall, 154; and Kakabekia, 167–68; and life span, 158; and photorespiration rate, 16
tendrils, 53, 54, 90
tentacles, in insectivorous plants, 52, 53
tension wood, in hardwood trees, 66
Thai Binh, and rice paddies, 19, 22
thallus, 113–14
theory of evolution, and Charles Darwin, 51
thermonuclear reaction, 5, 6–7, 8
thorns, as defense mechanism, 90, 91
TIBA, see 2,3,5- triiodobenzoic acid
tilth, 191
time-lapse photography, 37
timing mechanisms in plants, 74, 75, 80, 82
tip of plant, and tropism, 55–56
Ti plasmid, and pasteurization, 151
tires, and natural rubber, 177
tissue culture, 134; and aging, 159; in plant propagation, 119–20, 121, 124–25
tobacco, 124, 129
tomato: genetic experimentation, 137; propagation of, 124; and somatic hybridization, 129, 131
toxicity, 181, 183
toxic plants, 39, 42, 44–45; see also Astragalus; Xylorrhiza
tracheids, 26
trans-cinnamic acid, 178
transduction, 136

Index

transformation, 130, 136
transpiration, 25–26, 28, 29, 30
transplanting nuclei, and cloning, 142
transport cites in cell membrane, 35
transverse electrical potential, 63, 112, 113
tree, and growth direction, 65
tropistic responses, 56, 63, 66
Tumor Inducing Principle (TIP), 151
turgor, 26, 28, 78, 99
2,3,5- triiodobenzoic acid (TIBA), 66, 195, 196
2-4-5-trichlorophenoxyacetic acid (2-4-5 T), 198

undifferentiated plant tissue, 120, 127
unilateral light, and transverse electric potential, 112
urine saturated soil, 167
uterine implantation of blastomere, 141

vacuoles, 28, 92, 99
variable membrane permeability, 35–36, 37
Various Contrivances by Which Orchids are Fertilized by Insects, The (Charles Darwin), 51
vegetative growth, and inhibition, 195
vegetative process, in soybean, 80–81
vegetative propagation of water fern, 19
veins, and water movement, 26
velocity of light, 7
Venus's-flytrap (Dionaea), 90, 111
vesicle, and ingestion, 37, 130
vetch, 39

Vietnam, and defoliating chemicals, 196
virus: and cell fusion, 128; and cell membrane permeability, 128; and DNA, 136, 144; and ingestion by protoplast, 130; and seed transmittal, 134; and transduction, 136
virus-free plant strain, 134–35
vitamin, 14, 125, 159, 193; and inorganic versus organic gardening, 190–91

Wareing, Philip, 75, 77
water: and absorption in the water fern, 19; loss of (see transpiration); in photosynthesis, 3, 4, 13, 24, 80; plant use of, 24–30; in respiration, 14; and stomata, 28; xerophytes, and evaporation of, 30
water fern (Azolla), 18–23 passim
water fern–rice mutualism, 19–23
water-holding capacity of soil, 191
water impermeable layer, 26
water loss prevention, 26, 29
water soluble substances, 32, 34–35
water stress and abscisic acid, 29
water vapor, and open pore space, 28
water vapor loss, and abscisic acid, 78–79; from leaves, 25
waxy corn plant, and herbicide mutagenicity test, 183–84
waxy cuticle, and plant protection, 90
Went, Frits, 56
wheat, 16, 140
wild cucumber, 53, 54
wilting, 29, 78
winter, plant preparation for, 69–71
winter bud formation, 74, 75, 76
witches-broom, 149, 150
Wofatox, and Azolla protection, 23
woody aster, 39
woody shoots in winter, 76

Index

woody tissues, and water movement, 26
world population, and food, 9
wounds, and crown galls, 149
Wyoming State Board of Sheep Commissioners, 39

xerophytes, and arid conditions, 30
Xylorrhiza, 40, 41

yeast assay, 185